"Even as a conscientious consumer, obsessive recycler, and environmental advocate, it wasn't until I read *Plastic Purge* that I realized how little I knew about the ubiquity and consequences of plastics in my life. Thanks to SanClements's effortlessly casual style, not only did I enjoy reading about the 'good, bad, and ugly' ways that plastics have infiltrated my home, diet, and even my wallet, but I learned how to reduce my plastic footprint. This is the much-needed guide to living with the plastics you need and eliminating the ones that you don't." —Aron Ralston, *New York Times* bestselling author of *127 Hours: Between a Rock and a Hard Place*

"A fantastic guide to understanding plastics' role in an increasingly toxic world, including how to comfortably avoid prolific but unnecessary plastic products."
—Crissy Trask, author of *Go Green,*
Spend Less, Live Better

"We've never liked plastic, and *Plastic Purge* is a fantastic compilation of all of the reasons why we ought to seriously rethink our passion for plastic. SanClements provides clear, practical direction for anyone who wants to live a healthier, plastic-free life. We'll be following his advice closely."
—Bruce Lourie and Rick Smith,
authors of *Slow Death by Rubber Duck*

"*Plastic Purge* is a witty, thoughtful, and very useful guide for people looking for a way out of our collective addiction to dangerous, polluting, and (in many cases) completely unnecessary consumer products. SanClements offers a breezy, fascinating, and insightful look at our buying habits, and how we might change them in ways that will help ourselves and our planet." —McKay Jenkins, author of *What's Gotten Into Us: Staying Healthy in a Toxic World*

"*Plastic Purge* will make you smarter about the role of plastic in society and show you how to cut back on the presence of plastic in your own life. Read this book: You'll be healthier, and so will the planet. Though author Michael SanClements is an environmental scientist, he doesn't write like one. Which is to say *Plastic Purge* is lively, informative, and—dare I say about such a serious and important topic—a lot of fun." —Nena Baker, author of *The Body Toxic: How the Hazardous Chemistry of Everyday Things Threatens Our Health and Well-being*

PLASTIC PURGE

How to Use Less Plastic,
Eat Better, Keep Toxins Out of
Your Body, and Help Save
the Sea Turtles!

Michael SanClements

St. Martin's Griffin
New York

Library of Congress Cataloging-in-Publication Data is available upon request.

ISBN 978-1-250-02939-3 (trade paperback)
ISBN 978-1-4668-5260-0 (e-book)

St. Martin's Griffin books may be purchased for educational, business, or
promotional use. For information on bulk purchases, please contact
Macmillan Corporate and Premium Sales Department at 1-800-221-7945,
extension 5442, or write specialmarkets@macmillan.com.

First Edition: April 2014

10   9   8   7   6   5   4   3   2   1

For Hadley

CONTENTS

ACKNOWLEDGMENTS

First, I would like to thank Jasmine Faustino for contacting me and bringing this book to life—your guidance and friendship are truly appreciated! A huge thanks to Lindsay Edgecombe as well—I have enjoyed working with you tremendously and you are a fantastic agent. Also, thanks to Matt Martz for adopting the care of this book. Julia Suits made this book so much prettier with her wonderful illustrations. Many thanks as well to the good people at Grist for all they do and for giving me the opportunity to write about plastic in the first place. Much love and gratitude to Mary Perkins. Thanks for letting me sneak off for a year to write this book on top of a full-time job and never complaining about it. I owe you one. You are an amazing person and are going to be an amazing mother! Thanks to my family for always being there for me. You guys are the best and I love you so much.

The Laughing Goat coffee shop definitely deserves a mention, since I wrote nearly this entire book there. (Thanks for keeping me all hopped up on caffeine, Lauren.) Best coffee shop ever. Thanks to all my friends and my good dog Hank. Ella Levy, Basit Mustafa, and Sarah Elmendorf provided a lot of great banter that broke up the writing and made me laugh. Mary, Jesse, Ron, Ella, Sarah, and Dru provided input on the cover choice when I couldn't make up my mind. Lastly, a special nod to Iver Arnegard for "never being too wet to sell the river." Thanks, man.

PREFACE (AND HELLO)

"Humanity's plastic footprint is probably more dangerous than its carbon footprint," said Captain Charles Moore, the man who discovered the Great Pacific Garbage Patch in 1997. OK, well, maybe he was wrong. But . . . his statement still has something to teach us. First, there's clearly enough plastic in the middle of the ocean to freak Captain Charles right out, and second, it's illustrative of our ability to let massive global problems like oceans filling up with plastic debris (or climate change, the biggest challenge humanity has ever faced) dissolve into the background noise of our day-to-day lives. I mean truly, how many of us can say we've given real thought to something as pervasive in modern life as plastic? Probably not very many.

Until recently, that included me, too.

Yup, I admit that until the past couple of years, I, too,

hadn't given a tremendous amount of thought to society's and even my own plastic consumption, but now, well, I can't help but see it everywhere.

And since I'm in confession mode here, truth be told, I never intended to learn this much about plastic, and I certainly never intended to write a book about it. I'm writing this book largely as the result of a conversation I had in 2011 with some of the good folks at Grist.org, an informative (and snarky) environmental online publication. During that fateful conversation, I agreed to attempt to live without creating any plastic waste for two weeks. Prior to that conversation, I'd never even considered such a thing. Even though I am an environmental scientist, such a possibility had never crossed my mind. At that time, my comprehension of plastics and the degree to which they have infiltrated and affect our lives can be described only as incredibly shallow.

Looking back now, I'm blown away that the fact that our society is super-plasticized wasn't completely obvious to me. But once someone points it out to you, it becomes impossible to ignore. However, until that happens, it somehow manages to stay hidden without trying to hide. How does the saying go? The best place to hide something is out in plain sight. And that's precisely what plastic does—it hides out in the open right in front of us, getting into everything but rarely being noticed. Perhaps we can't see it anymore because it's always been a part of our lives, so we just take it for granted.

There were definitely a few things I vaguely understood about plastic that just about anyone knows. For example, I knew to an extent that disposable plastic shopping bags were not so great. I also knew bottled water was a bad idea, but I hardly understood the full story. The ocean gyres and turtles choking on plastic debris were not news to me. Yet I was mostly clueless about the relationship between plastics and human health. But once I started looking into it, I began to see the whole picture so much more clearly. And let me tell you, once you start to see what's going on, closing your eyes again becomes impossible.

So who cares? We've got plenty of problems. Why worry about plastic?

For starters, because plastic reminds me of an invasive species: it was introduced to solve a problem but did so with such enthusiasm that it created a host of worse ones. I'm referencing the fact that plastic was, to a degree, born from an attempt to find a replacement for ivory used in billiard balls so more elephants could keep their tusks, and plastic bags were created to save trees by replacing paper ones. Plastic isn't inherently bad, but if you get too much, or just a little of the wrong kind, in the wrong place, then you've got a problem. And it just so happens that we have this exact problem in the wrong place: Earth, the only home we have.

Reducing our plastic usage isn't an insurmountable problem, not even close. It just takes a little understanding and someone to guide you through the various tips and tricks that can help you eliminate unnecessary and even harmful plastic from your life.

This book is broadly divided into four different sections

designed to accomplish just that. The first part is a sweeping history of plastic that aims to explain when and why plastic became so ubiquitous in our lives. In the second part I cover some of the science behind our plastic usage. I address questions like: What resources are consumed to produce all our plastic? What are the human health and environmental effects of plastic? These are questions that have truly scary answers. The third part of the book lays down how I categorize plastic: the Good, the Bad, and the Ugly. Recognizing which category a plastic product fits into will help you figure out whether it's the type of plastic you want to keep around, avoid completely, or try and use less of. And the fourth part of this book is a detailed guide to help you reduce plastic consumption, keep toxins out of your body, and spare Mother Nature the excess waste.

Earth's population is currently right around seven billion, and should hit nine billion by 2050, which may seem like a really long time in the future, but isn't actually that far off. Regardless of whether or not you can comprehend the year 2050 or nine billion people, there is a simple truth inherent in these numbers. That truth is our planet is becoming a more crowded home with fewer resources. And unless we want increasingly worse resource shortages, we have to be increasingly diligent of how we elect to use those resources. Actually, we don't *have* to, we *need* to if we don't want our society to crumble, collapse, or morph into some very unpleasant place you'd rather not have your children inheriting. In this new reality, we are going to have to think hard about opportunities for conservation and seize them. We can't be squandering precious resources on unnecessary and inefficient products or practices. Any opportunity we have to lessen the pressures on our environment, infrastructure, and population must be acted upon. This is where plastic becomes relevant. Many aspects of our throwaway lifestyle are exemplified by and depend upon single-use plastics and the numerous cheap plastic products we regularly consume. It's a terribly inefficient use of resources with an enormous footprint, and we need to tighten up our use of this kind of plastic as soon as possible so we have room for the plastic we really do need: the kind that greatly improves our lives.

Here's where you come in. If you apply the knowledge you gain reading from this book, you will not only improve your own quality of life but also help make Earth a cleaner and better place to live!

INTRODUCTION

Plastic's versatility and functionality is what has made it so successful in modern society. It's this incredible versatility and range of characteristics that allows plastic to appear in so many products and serve so many uses. It's not much of an exaggeration to say you simply can't do anything today without coming into contact with plastic. But the proliferation of plastic is also driven in part by some random historical events, as well as the craftiness of a few individuals and companies who pulled off one of the greatest marketing campaigns in history.

In order to further discuss plastic, we need to get on the same page about its definition. The word "plastic" leaves a lot of room for interpretation. There are many different substances with a huge range of properties and characteristics that fit in this definition. However, here, in all its grandeur, is the definition of plastic according to the *Oxford English Dictionary*:

A synthetic material made from a wide range of organic polymers such as polyethylene, PVC, nylon, etc., that can be molded into shape while soft, and then set into a rigid or slightly elastic form.

You may have noticed that the word "polymer" is central to the definition of plastic. OK, so what's a polymer?

The American Chemistry Council Web site defines polymers as "compounds produced by polymerization, capable of being molded, extruded, cast into various shapes and films, or drawn into filaments and then used as textile fibers." Maybe you're wondering why I'm torturing you with words like "polymer" and "polymerization"? After all, when you picked up this book it wasn't because you wanted a new chemistry text. Sorry. But, there's just one more definition that I think we need to clarify if we're going to understand the history (and eventually the present and future) of plastics. According to the definition of plastic, *plastics are always polymers, but polymers aren't always plastics.* The first time I read that it didn't really resonate with me, but eventually the significance of that tongue-twisting statement became evident.

It's significant, especially in a historical context, because polymers share many characteristics with plastic but polymers occur naturally. In fact, by definition tortoise shells are polymers, as are cattle horns and rubber. So within the context of this book, when I say "plastic" or "plastics" I'm referring to fully synthetic human-made plastics. Unless I'm using "plastic" as an adjective, in which case I mean "capable of being molded or formed." Therefore, when I refer to a natural polymer or modern bioplastic, I'll specify which one I am talking about.

Now that you know what plastic is, it's time to open your eyes and discover just how plasticized your life really is!

It's nearly impossible to make it a week in our civilization without coming into contact with plastic. I know this, because I've tried. Seriously, put this book down for a second and take a look around you. What do you see? A cell phone? Laptop? Are you reading this on a tablet? Plastic, plastic, plastic! Not to mention the fact that plastic is in all kinds of things you wouldn't expect, from soup cans to cardboard milk containers to your disposable coffee cup.

I thought a good way to clearly illustrate my point would be to record every single (more or less) plastic thing I came into contact with during a typical day in my life. I made no effort to do anything out of the ordinary (for me) with regard to using or avoiding plastics. This was an interesting exercise (and one which I recommend everyone try out for a day).

Ready?

A Day in a Plastic Life

6:15 A.M.—Alarm goes off on my plastic cell phone, which is plugged into a plastic wall outlet.

6:18 A.M.—Flip plastic light switch in bathroom.

6:20 A.M.—Brush teeth with plastic toothbrush using toothpaste from a partly plastic toothpaste tube.

6:23 A.M.—Push aside plastic shower curtain. Turn on water, which flows through plastic pipes.

6:25 A.M.—Wash my hair with shampoo from plastic bottle.

6:45 A.M.—Eat cereal poured from plastic-lined bag and add milk from plastic milk jug. (Not to mention that my refrigerator has a lot of plastic components.)

6:55 A.M.—Pack lunch into Pyrex container with plastic lid.

7:00 A.M.—Place laptop into nylon work bag. Nylon is a plastic.

7:05 A.M.—Hop on my bike, which contains many plastic parts (like the seat and the grips), and bike to work.

7:15 A.M.—Use plastic key card to access building, walk across carpeting that contains plastic fibers. (Oh, by the way, your shoes likely contain some plastic.)

7:20 A.M.—Turn on computer and sit on chair containing plastic foam.

7:30 A.M.—Fill stainless steel reusable coffee cup from plastic coffeemaker in kitchen room. Screw plastic lid on top of reusable coffee cup. The coffee itself came from a plastic package.

10:00 A.M.—Print some papers with plastic printer. Staple them with plastic stapler.

10:40 A.M.—Recycle a piece of scrap paper I have been sketching an outline on by tossing it into a recycling bin made of plastic.

12:00 P.M.—Lunch with friends. In the lunchroom I see other people microwaving throwaway plastic containers in the plastic microwave, then sitting in plastic chairs at the tables to eat. *Is microwaving plastic bad?* I wonder (see page 100). I add salad dressing to my salad. The bottle is glass but the cap is plastic.

5:50 P.M.—Put on my fleece jacket (made from recycled plastic—thanks, Patagonia!), headlamp, running shoes, and running pants (which are also partially plastic! Nylon is the culprit here) and head out for a trail run with my dog, Hank. At some point on my run I have to pick up Hank's poop. This requires a plastic bag.

7:00 P.M.—Feed Hank. His food bag, while looking very papery, is actually lined with plastic.

7:45 P.M.—Grab a quick bite to eat. I get tacos that are sitting in a reusable plastic basket lined with wax paper, and the cup I was given for fountain soda was lined with plastic, not wax as most people usually think.

8:30 P.M.—At my favorite coffee shop, using my laptop to prepare for a talk I'm giving the next day, I order a tea. The tea bag is wrapped in plastic. There is a band playing live jazz at the coffee shop. I notice their instruments and speakers contain plastic. I am sitting in a wooden chair, but it has a thin cushion with a plastic cover. I get some water and pour it from a pitcher into the reusable plastic cups they have set out.

10:30 P.M.—I walk home from the coffee shop. The talk I give tomorrow will be projected via a plastic projector, via a plastic computer, powered by all sorts of plastic-coated wires.

What's your day like? How much plastic do you come into contact with?

It hasn't always been this way. Your great-grandparents didn't bring produce home from the store in plastic bags or keep leftovers in Tupperware. But today we are producing nearly three hundred million metric tons of plastic globally each year, only about 10 percent of which is recycled! So what gives? How did we go from no plastic to being so inundated with the stuff that we quite literally cannot escape it? As I said earlier, in the summer of 2011 my partner Mary and I tried to purchase no plastic and create no plastic waste for two weeks. When I blogged about it on Grist.com, a great environmental site that aims to spread awareness of environmental issues without boring you to death, I wasn't prepared for the response our little experiment would generate. Readers were hungry for information and wrote about the ways that our experiment inspired them to cut back on plastic, and even crowdsourced great tips.

But here's what else surprised me: in those two weeks, Mary and I felt healthier. We couldn't really eat junk food, since all of it is packaged in plastic, and we lost weight. We were already pretty healthy eaters before, but we both lost a few additional pounds as a result of attempting to cut out plastic. We had less *stuff* around the house and fridge, which also made us feel more organized and sane. It felt good to know we weren't sending crap into landfills or putting chemicals like BPA and phthalates into our bloodstreams (more on those in a moment). Turns out, cutting back on plastic simultaneously worked to help us lead healthier lives.

For the record, it's not possible to totally eliminate plastic from your life, and that isn't what this book is proposing. (In fact, we failed during our two weeks!) This book will show you how to reduce the waste and health risks from the plastic in your life while still using some plastic, which can be incredibly useful.

Wait, you say, risks? What risks? I'm hardly about to get tangled up in a six-pack ring and drown. Can something we use so regularly really be so bad? Plus, I really like my smartphone and my computer, and frankly, Tupperware is pretty handy, so I'm just gonna go ahead and toss my plastic bottle in the recycle bin, and problem solved!

Oh . . . how I wish it were it that easy! But alas, it's not. The story of plastic in our lives is far more complicated than simply using it, recycling it, and calling it good. As a matter of fact, the more I've looked into it, the more stunned I've become by what I have found. When you start to dig into the effects of our plastic consumption, they go far beyond the obvious and visible. There are environmental and health-related effects that you've never considered, and there is plastic in places you never expected it to be. For instance, did you know that:

♻ A study published in *The Journal of the American Medical Association* demonstrated that after eating soup from a can, the toxic chemical Bisphenol A (commonly referred to as BPA) was found to be readily detectable in urine? And by readily detectable, I mean "among the most extreme reported in a non-occupational setting." Now you may be wondering why I'm talking about BPA, a common ingredient in plastics, when you thought I was talking about canned soup? That's because most canned foods are lined with plastic. Surprised? Yeah, I was too.

♻ BPA levels in urine were recently linked to obesity in a study published in *The Journal of Clinical Endocrinology & Metabolism*. The study concluded that

BPA was positively associated with generalized obesity, abdominal obesity, and insulin resistance in those over forty years of age.

♻ Phthalates are another common class of chemicals found in plastics that have been linked to a host of health problems, particularly reproductive ones, including earlier breast development and possible increased breast cancer risk in females, reduced sperm count in males, and increased genital abnormalities in boys born to mothers with high exposure to phthalates. Yikes!

But BPA and phthalates are just the beginning. Unfortunately, there's a lot we just don't know about plastic. Scientists really don't have a handle on what exposure levels of these chemicals are toxic. Meaning they don't know how much is too much. Many of you have probably seen BPA-free products like water bottles, but what does BPA free really mean? Good question.

Some evidence suggests that BPA substitutes like BPS may be even more dangerous. BPS isn't the only substitute with dubious characteristics, as *The New York Times* reports: "Bisphenol B and Bisphenol F are other variants used instead of BPA in various products. In the limited testing done on those chemicals in other countries, scientists found Bisphenol B to be more potent than BPA in stimulating breast cancer cells." And this isn't the only instance where the unknown might be worse than the known.

Until now, there hasn't been one place where the average person can turn to really get a sense of what plastic means and has meant to us. This book will be that resource, and I hope that by the time you finish reading it you will fully understand the story of plastic from history to toxicity—it's all here. On top of that, this book serves as a prescriptive guide to teach you how to go about cutting back on plastic in all aspects of your life in order to be healthier and make the world cleaner while you're at it. And while I am a scientist and regularly deal with chemistry in my day job, you're not going to be buried in the chemical formulas for various plastics here. There are plenty of places to turn for that kind of information, and you can find a list of those in the resources and references section should you want to dig into the details.

After my experiment in living without creating any plastic waste, I realized that plastic falls into three categories: the Good, the Bad, and the Ugly.

The Good is your phone, camera, computer, medical equipment, ski bindings, etc. These things last a long time, and using plastic makes possible, or greatly improves, their performance. For example, would you ever want to backpack with a canvas tent? I think most people would opt for nylon to keep out rainwater. It's plastic like this that we don't want to get rid of, because this plastic creates more serviceable products that allow us to do things far more efficiently and, in many cases, do things that would not be possible otherwise. Think about medical and scientific equipment—much of it is plastic and isn't reusable and probably creates a lot of waste, but using these plastic products is extremely important for sanitary reasons. So instead of focusing on eliminating all plastic, we should concentrate on eliminating plastics that are unsafe or unnecessary, or what I referred to above as the Bad and the Ugly.

The Bad is stuff like plastic food storage containers or that new shower curtain you just hung up. It gets reused over and over, but while you are using it, who knows whether or not it's leaching nasty chemicals into your food or onto your body? It's nice to know that in some instances, like water bottles, there is a movement toward BPA-free plastics, which are now widely available for purchase, but even these may still not be safe. Once you start to ponder and read up on the bad plastic, it doesn't take long to realize that you are surrounded by plastics loaded with chemicals you most likely know nothing about, and that there are probably one thousand different pathways for these to enter your body, since there are so many instances where we are exposed to these products! Just think of baby products—are all the plastics in these safe? Let me answer that for you—not really. Later in this book I will dive much deeper into these topics, in both the sections on toxins in plastics (page 93) and in the more specific prescriptive sections in the second half of the book (page 117) that cover how to cut back on plastic and avoid toxins in products from baby bottles to sports equipment.

These are just a few of many more examples. But don't get

overwhelmed. Lucky for you, I've done the research and will provide you with the information you need to make the best consumer choices.

The Ugly is what I think of as lazy plastic: single-use plastic that's easily avoidable with almost no effort required to find a substitute. It's also the type of plastic you tend to see littering streets. Plastic grocery bags are the king of ugly plastic; they get used once and then they're done. Luckily, you've probably noticed that there's a growing awareness about these bags, and people are starting to bring their own reusable bags, and some groceries and cities are even banning plastic bags or offering incentives to encourage the use of reusable bags. Plastic bags are just the tip of the iceberg, and this book will show you what else you can do to get rid of ugly plastic in your life. And I promise you, it's easy.

Good, Bad, Ugly. That's the way I see it, and when you're done with this book, hopefully you'll look around and see it that way, too.

PART ONE

A Brief History of the Brief History of Plastic

"I just want to say one word to you. Just one word . . .
Plastics . . . There's a great future in plastics."
—ADVICE GIVEN TO DUSTIN HOFFMAN'S CHARACTER
IN *THE GRADUATE*

1.

IN THE BEGINNING

When I began researching the history of plastic, I didn't have a lot of grand expectations. Honestly, I wasn't really expecting to find anything that interesting. I didn't think it would be particularly boring either, I just thought I'd be reading a relatively straightforward timeline of progressively better plastics being invented over time. In retrospect, I should have known better. Science, and life, never work that way; there are always interesting and bizarre backstories.

I recognized a shift had occurred somewhere in the past moving us from a period of traditional materials like glass, metal, and ceramic to one in which plastic dominated. At some point we entered the Plastic Age. Yet I never fully considered why, when, or how this happened. When and why did nylon become popular? Or when and why did food cans become lined with plastic? And why were sturdy metal and wooden toys swapped for plastic ones? And why in God's name are plastic bags everywhere? Since things tend to make more sense when we know the backstory, I began to research and to read all about plastic. I wanted to know why, when, and how the planet became so plasticized that writing this books is even a possibility.

Before we try to understand exactly what happened here, it may be instructive to focus on where we are now, and how extremely full of plastic society is. It's tremendously difficult to convey the ubiquity of plastic in our lives, but I think this paragraph from the Society of the Plastics Industry Web site does so in a remarkably concise manner (www.plasticsindustry .org). I should also add that if I'm doing my job well, this little paragraph should hold a different meaning if you were to read it at the conclusion of this book. Anyway, here's what they have to say:

> Plastics have developed an amazing presence in our lives. From the most commonplace tasks to our most unusual needs, plastics increasingly have provided the performance in products that consumers want. In fact, if you woke up tomorrow and there were no plastics, you would be in for quite a shock. Life would be much more expensive and much less comfortable. And many of the conveniences you had come to take for granted would be gone. Mostly, though, you would be

surprised at the many products that had vanished—
things you had never thought of as being plastic.

There was a time when that passage wouldn't have meant
much to me. I'd likely have read it and thought "OK, sure,"
and carried on about my business. But, now I see the entire
story of our relationship with plastics tucked neatly within
that paragraph. In the two weeks I spent without purchasing
anything plastic or generating any plastic waste I gained a
new sense for the enormous role plastic plays in our daily life.
One thing's for sure—we most certainly take it for granted.
If you woke up tomorrow and there was no plastic, not only
would you be shocked, you'd also wake up on the floor because your mattress and box spring likely contain plastic
components.

Reading that statement from the Society of the Plastics
Industry also poses an excellent question: If I traveled back in
time, how far back would I need to go in order to wake up in
a world where plastic isn't a part of my daily life?

What's incredible is that you don't have to go very far back
in history to find yourself in a society that is free of plastic, or
at least so free of plastic you wouldn't recognize it.

The invention of plastic sort of occurred in the year 1862,
when a British man by the name of Alexander Parkes de-
buted a little something he called Parkesine at the International Exhibition in London. The reason I say that plastic was
sort of invented in 1862 is because Parkesine was not a fully

synthetic polymer, so it wasn't a fully synthetic plastic. Rather, it was made by mixing cellulose, an organic polymer, and also the major component in plant cell walls (it's what makes them rigid), nitric acid, and a solvent (basically a substance that dissolves something). The mixture could be heated, molded, and cooled into various shapes and was the first polymer producible in different colors without the application of dyes or a surface finish. Although this sounds pretty commonplace now, it was an enormous deal at the time, and Parkes seemed to know it. He did a fine job singing the praises of his new material in the leaflet he created for display along with some specimens at the International Exhibition. Parkes does a better job selling his product than I ever could, and his flowery language perfectly describes just how fantastically wondrously amazing Parkesine was for the time:

> A new material . . . now exhibited for the first time . . . the numerous purposes for which it may be applied, such as Medallions, Combs, Knife Handles, Boxes, Pens, Penholders. It can be made Hard as Ivory, Transparent or Opaque, of any degree of Flexibility, and is also Waterproof; may be of most Brilliant Colours, and has stood exposure to the atmosphere for years without change or decomposition.

Parkes wasn't the only one who thought his new material was something special. In fact, Parkesine ended up fetching the medal for "excellence in product" at the 1862 International

Exhibition and is considered by many to be the first plastic. It represented a groundbreaking advance in polymer science and a major expansion in the role people saw for polymers in society and industry.

Parkes had another telling thing to say about Parkesine, that it was "partaking in a large degree of the properties of ivory, tortoiseshell, horn, hardwood, India rubber, gutta percha, etc., and which will, to a considerable extent, replace such materials . . ."

Prior to the invention of Parkesine (and for some time after) many of plastic's current roles were filled by natural polymers like ivory, tortoiseshell, and the other materials Parkes mentioned in the above quote. In fact, there was a whole profession centered on boiling, grinding, flattening, cutting, and molding things like cattle horns, hooves, and tortoise shells into everyday items like combs, spoons, and lantern windows. "Horners" were the people who made horns into other stuff by peeling the horns along natural growth lines and making thin sheets of horn that could then be made into those lantern windows, spoons, and combs by layering sheets together until a desired thickness was reached and then molding and cutting them to shape. Because hooves, cattle horns, and tortoise shells are natural polymers, they are by definition "capable of being molded, extruded, cast into various shapes and films, or drawn into filaments and then used as textile fibers." They were also strong and lightweight and didn't shatter, but these natural polymers had their own set of problems, mostly deriving from the fact that these materials aren't uniform, which made production of each individual item a painstaking process. Massachusetts comb makers did invent some automated machinery to help ease the process, but it failed because each individual cattle horn had a different thickness, consistency, and flexibility. Machine settings might work for one piece of horn, but not necessarily work on the next, which resulted in a lot of broken combs and broken machinery.

Horns, tortoise shells, and hooves weren't the only natural polymers regularly used in roles filled by plastics today. There are other even more bizarre examples, like shellac. Most people think of shellac as a coating you paint on wood to keep it from rotting (which it is), but it's also a natural polymer. It was first used in India and eventually brought to Europe in 1290 by Marco Polo after his travels through Asia. Shellac comes from the insect *Kerria lacca* (also known as the lac bug). Specifically, shellac comes from its butt in the form of a secretion, which is also called lac. Lac is used as a coating for candies, like jelly beans, and in cosmetics, like lipsticks. And while eating or applying secretions from the rear end of an insect to your lips might strike you as disgusting, it's benign, and far less disgusting and disturbing than the chemicals modern plastics are capable of leaching into your food and drink.

Shellac, however, wasn't the easiest thing to harvest and produce. It needed to be scraped and separated from the bark and branches of the tree where the female lac bug eats and continually secretes the stuff, forming a tunnel, or cocoon-like structure. It can then be mixed with alcohol to form liquid shellac, which is applied to surfaces as a shiny protective coating. Solid shellac can be heated, molded, and used in making buttons, nobs, and phonograph records. Despite its range of uses and persistence into the modern era, shellac had its shortcomings. As was true of cattle horns and tortoise shells, the problems were related to a lack of uniformity and variations in color over time. Uniformity and durability were the ultimate goals when it came to plastic.

Despite not being a fully synthetic plastic, Parkesine overcame the uniformity issues associated with variations in the quantity and quality of traditional natural polymers. And while this was a major step forward in plastic production, Parkesine didn't hang around on the plastics scene for very long. The Parkesine company bombed in 1866 due to poor product quality resulting from Parkes's efforts to keep the costs of Parkesine artificially lower than the cost of rubber, another natural polymer.

Rubber, a natural latex, comes from trees. Specifically from the rubber tree, *Hevea brasiliensis*. Natural rubber is collected from trees by tapping them and collecting it as it runs out, much like maple syrup. In 1823 Charles Mackintosh received a patent for waterproof clothing using natural rubber. Unfortunately, natural rubber is super sensitive to temperature change, and the jackets basically melted when it was hot and turned incredibly stiff when it was cold. However, with a just a little tweaking through a process called vulcanization (the addition of little sulfur), natural rubber can overcome many of its shortcomings. Vulcanization was discovered by

Charles Goodyear in 1839 and greatly expanded the functionality of rubber. But natural rubber, vulcanized or not, has its limitations, especially with respect to color, as it comes only in black or very dark brown and therefore can't mimic or replace the tortoiseshell and horn products that everyone was used to. Do you see the theme here? This is yet another instance where a natural polymer is good, but problems with production, color choices, or uniformity prevented industry from taking it to a level of mass production across many products. So even with the vulcanization of rubber, industry would remain on the hunt for a consistent and cost-effective polymer to perform well in new products and as a substitute for difficult-to-process natural ones. Especially for use in billiard balls.

While his company may have bombed and Parkesine didn't turn out to be the magical product everyone was in search of, Parkes did succeed in a sense, because Parkesine kicked off the plastic party that's still raging today.

The first person to show up to that party was a guy named John Wesley Hyatt. Hyatt (and others, I'm sure) were driven, at least in part, by a ten-thousand-dollar purse offered by a New York City billiards company in the 1860s. The company was interested in finding a synthetic substitute for billiard balls, which at the time were made from ivory. The source of ivory, as you know, is elephant tusks, of which there is not an infinite supply. Hyatt pursued a new material, eventually inventing and patenting celluloid in April of 1869. His patent focused directly on the production of a new billiard ball, stating:

The object of my invention is to produce a composition which is adapted for being molded into a variety of useful forms . . . will serve as a good substitute for ivory in the manufacture of billiard-balls, and balls or articles of various descriptions, hardness and elasticity . . .

Celluloid, a mixture of nitrocellulose, camphor, and alcohol, was very similar to Parkesine, and Hyatt even credited Parkes during a lecture in 1914, saying that celluloid was not truly his own invention.

Despite filing the patent and successfully creating a synthetic billiard ball, it seems that Hyatt was never awarded his ten-thousand-dollar prize by the billiards company. The closest thing I have found to an explanation for this appeared in the March 17, 1940, edition of *The Spokesman Review,* and it wasn't much of an explanation at all, reading, "What happened to the $10,000 prize no one seems to know. In the long struggle to perfect celluloid the prize was apparently overlooked." Prize or no prize, Hyatt had high hopes for celluloid. He started both billiard and denture companies hoping to use celluloid as the main ingredient, but unfortunately it didn't work well in either instance, as celluloid dentures tasted terrible and the billiard balls performed poorly and couldn't mimic the elasticity of ivory (perhaps another reason he didn't get his prize money). But celluloid made great film! In fact, it made the early silent films possible and remained the dominant film type until it was replaced by safety film in the 1930s—since it turns out that celluloid had one little problem: it sometimes burst into flame and was ultimately responsible for many theater fires and fatalities.

The first truly synthetic plastic didn't appear until 1907, when Leo Henricus Arthur Baekeland invented Bakelite in his workshop in Yonkers, New York. *Time* magazine said Bakelite was "born of fire and mystery" and "the plastic of a thousand uses." Unlike the earlier modified natural polymers, Bakelite was resistant to heat, made an excellent insulator, was lightweight and easily moldable, and could be produced in a bunch of different colors. And guess what else? It made pretty good billiard balls! Bakelite quickly dominated the plastics market and appeared in everything from phones to billiard balls and kitchenware. Turns out *Time* magazine was pretty much accurate when they called it the "plastic of a thousand uses."

However, while Bakelite was incredibly popular, its production and usage was a small fraction of the percentage of plastic used today. In fact, prior to the end of World War II people outside of the military didn't really use plastic. Today annual global plastic production is about 165 times greater than it was in 1950.

In 1950, you didn't put your food in Tupperware, eat with plastic spoons, drink out of plastic cups, or give your kids LEGOs. Kids played with wooden blocks, sticks, rocks, or maybe balls. So what happened? Why at the end of World War II did everyone go plastic crazy?

During World War II everyone was cranking out wartime plastics, literally millions of pounds of this stuff, partly due to a couple of things—one being that the war itself led the military to greatly increase its production of plastics from 213 million pounds in 1939 to 818 million pounds in 1945. As a result plastic found its way into all sorts of products, revealing its many snazzy uses, like contributing to the successful manufacture of the first atomic bomb. Teflon (yes, it's a plastic) was critical in providing corrosive-resistant materials to handle the volatile chemicals needed in production of the A-bomb during the Manhattan Project. Depending upon what you read and how you think about it, you may even get the sense plastics were responsible for the Allied victory in World War II. Then the war ended. But damn, there were still all these companies producing plastic and all this plastic-production capability. So what to do?

You know who knew what to do? A group of people by the name of the Society of Plastics Engineers (SPE). First convened in 1942 to "provide and promote the scientific and engineering knowledge relating to plastics," these folks were totally bummed to see all that plastic potential get tossed by the wayside, so they stepped in to save the day. But they had a major and really interesting hurdle to overcome, which was that plastic products were seen as cheap crappy imitations of products made of traditional materials such as glass, metal, and wood. If you're old enough, perhaps you remember those solid metal Tonka trucks from your youth. Those things were super sturdy—nothing like the throwaway plastic toys we so often encounter today. People liked sturdy things then; I think people still like sturdy things today, too, but I think we've grown accustomed to a buy, break, and replace type of consumerism.

In the middle of the twentieth century, consumers weren't yet used to the throwaway lifestyle, so the SPE had their work cut out for them. They had to change the public perception of plastic from cheap and crappy to futuristic and marvelous. And they had to do so in a society that didn't live to consume and throw out products. The SPE focused on a marketing campaign that would wow Americans. Boy, did they really go all out when they held the first National Plastics Exposition. And you've got to hand it to them, because their strategy totally worked. The first National Plastics Exposition was held in Grand Central Plaza in New York City in 1946 and featured a whopping 87,000 attendees. The exposition kicked off with a demonstration by the US Navy showcasing their new methods for sealing machinery and weapons with plastic prior to storage. In addition to the Navy, over two hundred other exhibitors were present to highlight their new plastic creations. People were extremely excited by these displays. In fact, so many people attended the conference that it turned into a total fiasco, with long lines resulting in a pushy and unhappy crowd, which, coupled with extreme overcrowding, had the city threatening to shut the event down. I suppose that is a success of sorts. Ever since, the National Plastics Exposition has been closed to the public.

This wasn't the only grand marketing scheme designed to get everyone jazzed on a plastic future. Ever heard of the company Monsanto? I'm sure you have. Monsanto is the world's largest biotechnology company and a producer of products like DDT, Agent Orange, and Roundup—a widely used and highly controversial pesticide—and genetically modified seed. Interestingly, Monsanto is also one of the companies responsible for convincing the public that we all need to have plastic everything all the time. More, more, more! While the SPE brought the National Plastics Exposition to the table, Monsanto had their own over-the-top plastic marketing scheme in mind.

In 1957 they decided to construct "The Home of the Future" on a little plot of land in Anaheim, California, not all that far from Disneyland. The Home of the Future was made entirely of plastic and looked very similar to the Jetsons' house. In the case of Monsanto's Home of the Future, the future looked a lot like a typical home in 1987. The home made some pretty amazing predictions about modern life. For one, the house contained a microwave oven as the kitchen's showpiece—an impressive and not so off-the-mark prediction for the time. The home was constructed entirely of plastic, and I mean entirely: plastic ceilings, walls, doors, floors, furniture, and cutlery. And while the modern household isn't all plastic all the time, Monsanto's predictions weren't terribly inaccurate.

Plastics found their way into the daily lives of Americans (and citizens of other countries, too) in countless ways during the mid-twentieth century. Sometimes new plastics resulted in minor changes or the arrival of pleasant new conveniences, while in other instances, the introduction of plastic products had profound societal implications, like helping the Allied Forces to emerge victorious from World War II by contributing to the mass production of parachutes, aircraft parts, helmet liners, bazooka barrels, the A-bomb (as I mentioned), and more.

Discussing all the ways new plastic products affected American culture is a project unto itself, but we can't call this part "A Brief History of the Brief History of Plastics" without diving into a couple of these stories. Here are my two favorites, which are the stories of nylon and Tupperware. They do a wonderful job of bookending the sweeping changes and downright insanity accompanying America's blossoming relationship with plastic. Plus, they are both pretty entertaining and even downright weird stories.

2.

A STORY ABOUT NYLON

From Coal, Water, and Air to War and Riots!

Wallace Hume Carothers, an employee of DuPont, an American chemical company founded in July 1802 as a gunpowder mill, invented nylon in February 1935. Nylon was the first synthetic textile fiber and, seventy years later, is still used every day. Sadly, Wallace Carothers never got the chance to see nylon become such a hit. He had a long history of battling depression and committed suicide the year after its invention. Weirdly, the *Washington News* ran a story around the time of Carothers's death proclaiming that nylon was made from cadaverine, a nasty chemical that forms during the decomposition of corpses. This isn't completely false. Nylon can be made using cadaverine, but it's not. It turns out, cadaverine can also be extracted from a resin of coal, which is where the ingredient used in nylon orginates. Nonetheless, extracting chemicals from coal is a far less remarkable story than extracting chemicals from corpses—and it seems people often remember the crazier or more extravagant story. As a result, nylon suffered a setback in its reputation, but one it ultimately overcame.

Similar to the earlier hunt for an ivory replacement, the creation of nylon was also driven by the search for an alternative to a natural product, in this case silk. Specifically, silk for women's stockings. In 1930, prior to the invention of nylon, American women were purchasing around eight pairs of stockings each, adding up to a total cost of seventy million dollars annually, and all that money was being exported overseas to Japan, the producer of all that silk. This transfer of funds certainly helped motivate the search for an American-made replacement. While we take nylon for granted today, its invention was completely revolutionary. It's not an easy thing to wrap one's head around—to imagine a world without nylon that suddenly became a world with nylon. But it was a big deal, guys! People absolutely freaked out. In fact, from the second nylon was announced to the public on October 27, 1938, the entire nation went nuts. Even *The New York Times* joined in the excitement, with headlines proclaiming "New Hosiery Strong as Steel" and an editorial titled "Time Defying Hosiery" that ran on October 29, the day before Orson Welles terrified the nation with his reading of *The War of the Worlds*. I find this superinteresting considering that at the time these articles were published just about nobody had seen or used nylon before. Despite all the fanfare, it wasn't until the following year, on October 24, 1939, that nylons became available in a first release to the wives of DuPont employees, who frantically scooped up all four thousand available pairs in three hours. Nylon then made its way into other products, products it still appears in today, like toothbrushes and fishing line. Once nylon made its way into a few publicly available products, it was off and running, and its growing public presence didn't slow down until the start of World War II—but I'll get to that in a second.

First, we need to talk about Nylon Day, May 16, 1940. Nylon Day was the first public release of nylon stockings and was quite the event. On Nylon Day, four million pairs of nylons hit the store shelves, and two days later all of them were gone. Poof. At $1.15 a pop, which in 1940 was equivalent to about $18.91 a pair, this was an astonishing sales success! In 1941, over one hundred million pairs of nylons were sold. However, once the war started, all that nylon production got sucked into the war effort. Instead of stockings, nylon was used to make parachutes, ropes, and hammocks. At the same time, getting your hands on silk in the United States had become nearly impossible, due to the war with the Japanese and all. This caused quite a conundrum. With both silk and nylon stockings unavailable, people did what people tend to do: they adapted. In this case, they adapted by painting stockings on their legs with skin-colored makeup and drawing a line up the back to imitate the seam. This wasn't something only a few people desperate for stockings did, it became downright common practice. Liquid stockings became a regular part of American culture. Eventually, and thankfully, World War II ended and the U.S. nylon supply was once again free to be used in the production of stockings. DuPont didn't dawdle. Only eight days after the Japanese surrendered, DuPont announced they would be returning to the production of nylon stockings. By this point, just about everyone and her mother was craving some nylons and DuPont promised to deliver some 360 million pairs—enough so that every American woman was able to bring their stocking supply back up to a respectable number. Unfortunately though, this turned out to be wishful thinking, and when the nylons hit the stores there weren't nearly as many as promised and demand didn't decline with supply. Women all over the country began fighting for nylons. Rioting, actually. According to some newspaper headlines of the time: "Nylon Mob 40,000 Strong Shrieks and Sways for Miles!" "Women Risk Life and Limb in Bitter Battle for Nylons," "Nylon Sale No Casualties." Crazytown, right? There's actually a pretty iconic photo of a woman sitting on the curb in the midst of the city putting on her new pair of nylons. She's looking contentedly at her leg while a huge line of customers stands behind her. She'd clearly just scored a pair and parked her butt on the curb immediately outside the store, unable to wait even a minute longer to don them.

I think this little story serves both as a rather unique piece of American cultural history and as an illustration of the incredible convenience that plastic products bring to society in the form of durable, easily producible goods that are free from the constraints of natural products like horn and silk. The truth is, plastic tends to perform as well or better than the original substance it's replaced. It performs so well that people will line up forty thousand deep and literally fight for it when denied its convenience.

Nylon isn't the only example of a plastic product whose invention rippled through society in interesting and unexpected ways. Tupperware's invention had big implications for American culture and American consumerism, too. As with nylon, people got what might be described as weirdly over-enthusiastic about its invention.

3.

TUPPERWARE

A Classic Plastic Product We're All Familiar With

Tupperware has an interesting story, too. Much like revolutionary ideas, revolutionary plastic products were met with a bit of resistance upon initial release. Most people didn't like or just plain ignored Tupperware when it first came to market. But fast-forward a few years and people went freaking crazy and wanted to buy it all. Tupperware may have a permanent place in the minds, refrigerators, and even museums of America (thanks to the Smithsonian), but it totally flopped when it was first invented. It wasn't until

Brownie Wise, a brilliant and pioneering American business-woman, came to the rescue that Tupperware would become popular. In the end, though, Brownie would get totally screwed by Tupperware's inventor, and this whole saga would play out like a Greek tragedy—an unfortunate fate for an incredible woman who broke a lot of ground.

Earl Silas Tupper, the inventor of Tupperware, was a pretty intense guy and was absolutely determined to become a millionaire. He was also, it turned out, kind of a dirtbag.

Earl worked as a tree surgeon in Massachusetts by day and spent his evenings filling up notebooks full of inventions—some which were absolutely ridiculous, like the fish boat, which was basically a harness that allowed a really big fish to be attached to the bottom of a boat to propel it around. Eventually, however, he started his own company, Tupper Plastics. In his continued quest to strike it rich, Earl convinced DuPont to give him their polyethylene slag—a by-product of the oil-refining process that DuPont thought was worthless. Tupper experimented tirelessly with the stuff.

It's important to recall that at the time Earl Tupper was experimenting with his slag, plastic was not used for food storage. Plastics weren't that popular yet and were still considered brittle, smelly, and greasy. Plus, developing handy containers for prolonged food storage wasn't really an issue considering the fact that most people didn't have refrigerators yet. Nevertheless, Tupper used his leftover slag to create a nonsmelly, nonbrittle plastic storage container that he called the Wonder Bowl. While a usable plastic bowl in itself was an amazing feat for the time, it was the air- and liquid-tight top that made Tupper's invention truly spectacular. For the first time ever, we had a durable, airtight, unbreakable container effective at storing food for a long time by sealing in moisture with the famous "burping seal." Tupper got the idea for the lid from the paint cans at his local hardware store, and simply swapped the seal to the outside of the container. Genius really, but as fantastic as all this was, it wouldn't guarantee the success of Tupperware. A good marketing campaign was needed as well.

When Tupperware was introduced to the world in 1946, not much happened. Sales were so poor it was eventually pulled from the shelves in 1951. However, there was one exception. Her name was Brownie Wise. Brownie, formerly a saleswoman for Stanley Home Products, had started selling Tupperware in 1949 and was making really good money at a time when just about every other avenue of sales was falling flat. Earl Tupper took notice of this when Brownie Wise called him to complain that her order was late again and suggested he start selling his product in people's homes. She explained to Earl that her method for moving massive amounts of Tupperware was to sell it in people's homes in a party-like atmosphere; this was a novel concept for the time, and the origin of the now-renowned Tupperware party, which we've all heard of. Brownie, being the adept saleswoman she was, intrigued Tupper with her pitch. He invited her to Massachusetts to learn more. Brownie always made the sale, and once she had Tupper's ear, he was powerless to say no. In 1951,

Brownie would be appointed general sales manager of Tupperware Home Parties.

The Tupperware party originated around 1950 and still exists today. "Exists" isn't actually the correct word, so allow me to rephrase: Tupperware parties are absolutely thriving today! Rick Going, CEO of Tupperware, estimates there's a Tupperware party every two seconds globally. They've even morphed into the selling other products these days, like solar panels, kitchen gadgets, or even adult toys at Passion Parties. There's a lot of beauty in the Tupperware party methodology. When it's a friend or neighbor selling something in a comfortable and familiar environment, people feel they are less likely to be swindled than they would if it was a stranger selling them something—especially if the person selling said product is both a friend and consumer of the product.

Rick Going even posits that Tupperware parties are the grandmother of social marketing. In support of this argument, he points toward Tupperware's spending zero revenue on advertising; it's all friend to friend.

In 1951, Tupperware had been pulled from store shelves and Brownie Wise and her sales force were in charge of selling it exclusively via parties. Brownie proved to be a marketing genius. By 1954, she was leading a multimillion-dollar corporation and had become the first woman to appear on the cover of *BusinessWeek*. At a time when women weren't typically executives, Brownie was destroying barriers in American culture. And not just that—Tupperware and its sales model played an enormous role in the financial independence of women during the 1950s, especially single mothers and divorcees, of whom Brownie Wise was one. Tupperware parties gave women the ability to earn their own money, on their own terms, quickly.

The period between Brownie's appointment as general sales manager and the peak of Brownie's career involved some pretty bizarre antics. As Tupperware boomed, Brownie amassed an army of able saleswomen and spearheaded some spectacular marketing campaigns. She organized Tupperware Jubilees—massive annual four-day themed parties at company headquarters in Florida. The parties were designed to build loyalty among the Tupperware ladies. Headquarters was not just your average corporate headquarters—it was a crazy, over-the-top, Tupperware-themed wonderland. According to an episode of *American Experience*: "[Brownie Wise] made up new traditions. She baptized Poly Pond with polyethylene pellets, showed her dealers how to place their wishes in two-ounce Tupperware containers, then toss them in the wishing well." At a Gold Rush–themed jubilee, six hundred shovels were laid out and Tupperware employees scrambled to dig up prizes like watches, television sets, diamonds, and mink coats! Jubilees weren't all fun and nonsense, though. Serious business was conducted as well. Motivational speeches, sales classes, and even tests and a graduation. Despite the lavishness of it all, the jubilees were no doubt effective in bringing the employees together and building morale. After all, Brownie was deeply loved by her employees and her influence would continue to grow.

By 1957, Brownie and her team were kicking so much ass that the company couldn't keep up with her orders. She confronted Earl about the bottleneck in production, which made him angry. From there on out, Brownie would begin to lose hold of the Tupperware empire she'd created. Tupper began to get annoyed with her expenditures, especially on things like the jubilees, but what really enraged Tupper was the publication of Brownie's autobiography, *Best Wishes, Brownie Wise,* which Earl viewed as self-aggrandizing publicity.

In the end, after all that Brownie did for Tupperware, she was fired. She was given a lousy onetime payment of thirty thousand dollars and forced to sign an agreement saying she wouldn't work for any of Tupperware's competitors. They even had all photos of her removed from the offices, basically erasing her history at the company. A year later, Tupper sold Tupperware to the Rexall Drug & Chemical Company for sixteen million dollars. Brownie tried to start her own cosmetics company, Cinderella, which folded after one year. She lost everything.

The patent for Tupperware expired in 1983, one year before Earl did.

From 1950, the year before Tupperware was pulled from shelves and put in the more competent hands of Brownie Wise, to 2011, plastic production has increased from 1.7 million tons to 280 million tons. And as per usual, we Americans are gobbling up more than our fair share. On a per capita basis, Americans consume more plastic than anyone else in the world. The average American consumed 326 pounds in 2010, versus the average Eastern European, whose per capita consumption is estimated at less than 53 pounds per person. So basically, we are consuming more than six times the amount of plastic our Eastern European friends across the pond are. In the interest of being fair and thorough, I want to take a moment and play devil's advocate. So let us contemplate this: maybe everyone in Eastern Europe is missing out? Maybe they don't have as much great stuff as we do and our plastic consumption is reflective of our super-fantastic lives, which everyone else aspires to and is just insanely jealous of?

My answer: I think this is a fairly common perception in America, a perception that is, at best, wishful thinking, and at worst, extremely harmful on both personal and political levels. Despite the fact that we may feel like we're number one at just about everything, we actually lag behind many nations in many ways. Our plastic consumption isn't representative of our great lifestyle; I'd argue rather that it's linked to our status of having worse health and shorter lives than people in other wealthy developed nations. Other factors certainly enter into play here, but cheap disposable plastics are symbolic of, and perhaps even directly linked to, the declining physical and mental health of American society.

So here we are, it's 2014, global population has surpassed seven billion, resources are becoming more scarce, and we're overrun with plastic. It's ironic that in a roundabout way plastics, which now contribute to so many environmental problems, stemmed from an effort to slaughter less elephants. A fine intention indeed, but one completely at odds with the current reality of plastic in the modern world.

Now that you know a little bit about the history of plastic and its rise in our society, in the next section we'll focus on the science and resource issues involving plastic.

PART TWO

The Science Behind the Plastic

4.

HOW IS PLASTIC PRODUCED?

n case you have forgotten the definition of a polymer, let me remind you that the American Chemistry Council Web site defines a polymer as "any of various complex organic compounds produced by polymerization, capable of being molded, extruded, cast into various shapes and films, or drawn into filaments and then used as textile fibers." This definition leaves room for polymers to be either synthetic or natural. As mentioned previously, an easy way to remember this is: All plastics are polymers, but not all polymers are plastics. And recalling our history, we saw how the search for a synthetic replacement to commonly used natural polymer sources such as cattle horns and elephant tusks spurred the plastic revolution. Once uniform mass-producible synthetics came along, usage of natural polymers, for the most part, fell by the wayside.

Today your average plastic item is made from fossil fuels,

mostly oil. Plants can also be used to make plastics—or bio-plastics, as they are commonly called. Bioplastics account for less than 1 percent of global plastic production. Nevertheless, we are going to talk about them in detail in chapter 8, because they offer a potentially promising but also complicated alternative to traditional plastics. But for now, let's focus on the footprint and production of the far more common fossil-fuel-based plastics.

If you want to make some plastic (traditional style, not bioplastics), the first thing you need to do is score some oil, coal, or natural gas. For the sake of this example, let's make ours with crude oil. Much like making booze, making plastic begins with a distillation process. Which, considering how drunk we as a society are on plastic, strikes me as fairly appropriate.

Why are we distilling oil? Well, the distillation process is part of the refining process. It's used to separate out the different fractions contained within the oil. Once separated, different parts are suitable for different uses. As you heat oil, certain compounds will turn to vapor at lower temperatures than others. As component A turns to vapor, it flows through a tall column where it can be cooled, returned to liquid form, isolated, and collected. The remaining crude oil can then be heated to even greater temperatures to capture fractions with higher boiling points. If the goal of all this heating and capturing is to make plastic, then the fraction we're most interested in is *naphtha*.

Naphtha is a volatile flammable hydrocarbon liquid very

similar to gasoline. As a matter of fact, it is the predecessor to high octane gasoline. And although it is distilled like liquor, you don't want to drink this stuff. Naphtha comes in two types: light and heavy. Light naphtha boils between 86 and 194 degrees Fahrenheit, while heavy naphtha can handle a bit more heat, with a boiling range of 194 to 428 degrees Fahrenheit. Light naptha is the crème de la crème when it comes to making plastic because of its greater paraffin content, which is central in making olefins.

Olefins are produced by further processing light naphtha through a technique called *cracking*, which uses steam heat to split naphtha into smaller hydrocarbon molecules (these being the olefins). Two of the olefins, ethylene and propylene, comprise the basic building blocks of most plastics.

In no particular order, some of the world's largest plastic producers are:

BASF, Germany
Dow Chemical, USA
INEOS Group, England
LyondellBasell, Netherlands
ExxonMobil, USA
SABIC, Saudi Arabia
DuPont, USA
Total, France
Formosa Plastics Group, Taiwan
Bayer, Germany

Ethylene and propylene need a little more love, or specifically polymerization, before the plastic production process is finished. Polymerization is the linking of molecules to form polymer chains. Literally hundreds of different types of plastics can be made by varying the linking and structure of these polymer chains. There's much more to come about the many types of plastic in later chapters, but for now, what you need to know is that plastics fall broadly into two categories: thermoplastics or thermosets.

Simply put, thermoplastics soften when heated and can be melted and molded over and over, while thermosets cannot and cure into nonmelting insoluble forms. The inability of thermosets to be melted and reshaped is driven by the chemical structure of the linkages that bond them together. Thermosets are comprised of cross-linked molecules that prevent them from being reshaped once formed.

It's not a particularly critical distinction, but I want to point out that many items I didn't realize were plastic are indeed thermosets. The foam inside your office chair, or possibly your couch, is likely polyurethane, a thermoset plastic. Weird, right? I admit to having had no idea the foam in a lot of furniture was a plastic. This is just one little example. There are many more instances of plastic popping up in the strangest of places that we'll touch on as you read through this book.

Following the process of polymerization, "raw" plastics in the form of pellets, powders, or granules are further processed into an infinite number of items using industrial processes with names like "extrusion blow molding" or "injection molding."

Different methods are applied depending on the plastic used and the product being manufactured.

Soda bottles are made using extrusion blow molding, a process vaguely similar to blowing glass, but using plastic and big industrial machinery. Other plastic items, like crates or chairs, are made using a technique called injection molding, which consists of basically pouring liquid plastic into a mold of the desired dimensions and shape.

Entire textbooks exist on the methods used to shape raw plastics into the goods we all use every day (*Industrial Plastics: Theory and Applications* is one), so I haven't covered them in detail here. I find the actual production of plastics to be a bit boring, but I wanted to provide you with an extremely basic understanding of the process.

5.

HOW MUCH FOSSIL FUEL IS CONSUMED IN THE PRODUCTION OF PLASTIC?

Plastic facts have a knack for being vague and elusive. One would think a central question like the quantity of fossil fuel used in plastic production would be easy to find an answer to. But this question isn't all that straightforward, and trying to find a solid answer is downright difficult. Umbra Fisk from Grist's Ask Umbra (an excellent Dear Abby–esque column for questions related to sustainability) says it quite well when she states that the amount of oil used in producing plastic is "one part of a complicated harvest." One way to address this question is to use an approach called life cycle analysis or life cycle assessment (LCA).

A Quick Overview of Life Cycle Analysis

Life cycle assessment and *life cycle analysis* are interchangeable terms both bearing the acronym LCA. According to the United States Environmental Protection Agency (EPA), LCA is a technique to "assess the environmental aspects and potential impacts associated with a product, process, or service." Sounds pretty handy, right? Well, it is, except there's a caveat: LCA must be done correctly, and unfortunately, there's a lot of room to cheat, fudge things, or lie by omission, which makes LCAs both complicated and simple. LCAs are also an excellent framework for critical thinking. Learning the basic principles will hopefully help you to think more clearly and inclusively about processes and products.

In theory an LCA is a "cradle to grave" assessment of the cost/impacts of all the inputs and releases associated with a product from its manufacture through its disposal. The analysis should include the cost of all environmental and societal impacts. The EPA lays out what an LCA should be doing in three simple bulleted items:

♻ Compiling an inventory of relevant energy and material inputs and environmental releases

♻ Evaluating the potential environmental impacts associated with identified inputs and releases

♻ Interpreting the results to help you make a more informed decision

For example, the life cycle analysis of a plastic bag should include costs associated with:

♻ Extracting oil, including the environmental impacts and energy used in extraction
♻ Producing the plastic, including the energy required, emissions released, and potential harmful side effects
♻ Transportation, including how much energy is used in moving bags from factory to stores, etc.
♻ Life expectancy of the product, including how long it will last or if it can be reused
♻ Disposal, including if the product can be recycled or if it ends up in a landfill. What are the environmental impacts of those choices?

You can see how this gets complex fast. You can also see how easy it would be to miss a step or two and come up with a completely different answer. Maybe someone forgets to factor in the cost required to produce the plastic. Or perhaps in an LCA for bioplastics you forget to factor in the energy required to plant, fertilize, harvest, and transport the corn used in producing it. That might change things considerably.

So remember, next time someone makes a claim about the costs associated with a product or process, ask yourself: are they accounting for everything?

The amount of oil used in plastic production varies depending upon where you draw the line or how thorough your LCA is. For instance, what appears to be the best estimate of global oil usage for plastic production is 4 percent. And if you were doing a less-than-thorough job with your LCA, you might leave it there. However, if you take into account the energy used in production (e.g., distilling, naphtha capturing, cracking, polymerizing, shaping, etc.) of the plastic and other things like electricity used at the factories, then we need to toss in another 4 percent, for a total of 8 percent of global oil production. According to the United States Energy Information Administration (EIA) that's an estimated 2,595,880,000 barrels of oil for all the plastic manufactured in 2012. One barrel holds quite a lot of oil. Forty-two gallons, to be exact. All this oil is allocated to producing the many different types of plastic and plastic products we encounter every day.

6

THE MANY TYPES
OF PLASTIC

What Do the Numbers Inside the
Little Recycling Symbols Actually Mean?

For me, and likely for you as well, the first thing that comes to mind when someone mentions types of plastic is that number inside a recycling sign. Technically, these symbols are part of a system developed in 1988 by the Society of the Plastics Industry (SPI) called the Resin Identification Symbol System (RISS). (Sometimes it's referred to as the Material Container Coding System.) When it was first developed, most plastics were types 1 through 6. There is now a category number 7, which got slapped with an "other" label and contains many new plastics that have come into existence since 1988.

These symbols give the impression that all seven categories of plastic are regularly and easily recycled. After all, the majority of the symbol is a recycling sign! Unfortunately, though, this isn't the reality. *Just because a plastic item has a RISS symbol doesn't mean it can be recycled.* Only plastics numbered 1 and 2, or PET (soda and water bottles) and HDPE (milk jugs and detergent bottles), are routinely and efficiently recycled. Plastic bags are typically plastic number 4, or LDPE, and are accepted for recycling at most grocery stores, but the actual recycling rates are abysmal. RISS codes serve as a key to the common types of plastic one might encounter in daily life. And while not all plastics with a RISS code can be recycled, the symbols are incredibly informative and do contain information related to recycling. Because once you're aware of the type of plastic you're dealing with, it can be checked against the guidelines provided by the recycling program in your town or city (which you will have to look into on your own). There are two good ways to find this information: by visiting the Recycling Locator on the Earth911 Web site at http://earth911.com/recycling or simply Googling the guidelines for where you live (e.g., "Boulder recycling guidelines" or "Cleveland recycling guidelines" should do the trick). If you live in a very small town, you may need to contact your town hall by phone. The Earth911 Web site is a great source that allows you to search by item or by zip code and compiles all the information you need for successful recycling.

Familiarizing yourself with the more common plastic types is critical to understanding the story of plastic and the myriad ways in which plastics are capable of affecting you, your loved ones, and the environment. I've tried to summarize much of this information for you in the "Quick Guide to Seven Common Plastics" text box on page 57. Just remember, this table serves as a guide; it is by no means all-inclusive. Many of these plastics may be found in numerous other places or not found in some. For example, not every salad dressing bottle is necessarily made from plastic number 1 (PET). My hope is that, rather than being completely inclusive, I can provide you with a reference and framework for you to work from. In the future, before tossing something into your recycling bin or trash, you'll want to check the RISS code to see exactly what you're throwing out and also to gain insight into what kinds of chemicals you might have ingested, via leaching, from that bottle or container.

A Quick Guide to the Seven Common Plastics

Number 1—Polyethylene Terephthalate (PET) or (PETE) *Common Uses:* Soda and water bottles, salad dressing bottles, fibers like polyester. *Safety Concerns*: Possible leaching of antimony and phthalates. *Recycling*: Commonly recyclable.

Number 2—High Density Polyethylene (HDPE) *Common Uses:* Milk jugs, juice bottles, most laundry and dishwashing detergent and household cleaner bottles, some shampoo bottles, some trash and shopping bags, butter and yogurt tubs, cereal box liners. *Safety Concerns:* Low risk of leaching. This is a relatively safe plastic. *Recycling:* Commonly recyclable.

Number 3—Polyvinyl Chloride (PVC) or (V) *Common Uses:* Cling wrap, shower curtains, rubber duckies, some laundry and dishwashing detergent bottles, some shampoo bottles, clear food packaging (including some meat and deli wraps; not Saran Wrap, which is now made of polyethylene), wire jacketing, medical equipment, siding, windows, and piping. *Safety Concerns:* Cooking food in PVC packaging releases endocrine disrupters; toxins are released upon burning. *Recycling:* PVC is rarely recycled.

Number 4—Low Density Polyethylene (LDPE) *Common Uses:* Squeeze bottles, bread bags, frozen food bags, dry cleaning bags, shopping bags, tote bags, clothing (more typically athletic wear), furniture, some carpets. *Safety Concerns:* Low risk of leaching. This is a relatively safe plastic. *Recycling:* Some recycling programs take LDPE—check your local guidelines; most grocery stores take plastic bags for recycling whether they be number 2 or number 4.

Number 5—Polypropylene (PP) *Common Uses:* Some bottles (like ketchup), reusable containers (like Tupperware), clothing (long underwear), ropes, some carpets, bottle caps, some plastic furniture. *Safety Concerns:* Low risk of leaching. This is a relatively safe plastic. *Recycling:* Some recycling programs take PP; check your local guidelines.

Number 6—Polystyrene (PS) *Common Uses*: Takeout food containers, disposable plates and cups, red plastic Solo cups, Styrofoam, meat packaging trays, some disposable coffee cups. *Safety Concerns:* Leaches toxins, including styrene. The probability of leaching increases with heating. *Recycling:* Rarely recycled, but some programs do; check your local guidelines.

Number 7—Other (OTHER) Number 7 represents a wide variety of plastics that don't fit into categories 1 through 6. These include polycarbonate and many others. *Common Uses:* Big water coolers, sippy cups, some reusable water bottles, sports equipment, medical and dental equipment. *Safety Concerns:* Polycarbonates leach BPA and should be avoided. *Recycling:* Rarely recycled, but some programs do; check your local guidelines.

Note: Statements regarding recycling refer to the availability of opportunities for recycling, not the amount of plastic actually recycled. For example, PET is easily recyclable, but less than 30 percent of PET plastic bottles and jars were recycled in the United States in 2011.

Tips for Effective Plastic Recycling

1. Familiarize yourself with local recycling guidelines. You can even print them out and hang a list near your recycling bin(s) or on your refrigerator.

2. If you're curious about "cartons" (e.g., juice boxes, soup boxes, almond milk boxes), check out the Carton Council Web site, www.recyclecartons.com, and remember that plastic is lurking where you least expect it.

3. Remember, despite the recycling symbol on all plastics, not all plastics are recyclable. Prior to purchasing plastic products, check the number on the container to ensure you can recycle them where you live. Avoid plastics that you cannot easily recycle.

4. If you have questions about recycling (can I recycle this bread bag?), you can also check the Earth 911 Recycling Locator (http://earth911.com/recycling) to find the closest place to recycle just about anything.

5. *When in doubt, throw it out!* Before you beat me up for being wasteful and a heretic, consider this: a stray bottle *can* ruin an entire batch of recycling.

7.

RECYCLING PLASTIC

The EPA has this really great tool, the Individual WAste Reduction Model (iWARM), that does a fantastic job of putting things in perspective. The iWARM model translates the energy savings from recycling things like plastic bottles, rather than trashing them, into easy-to-comprehend energy units like, for example, the hours that energy could power your laptop, TV, or a lightbulb. The model includes options for seeing the amount of energy derived from plastic bags, gallon milk jugs, plastic bottles (20-ounce and 2-liter), and detergent bottles. And it's not just limited to plastic—there are a slew of other recyclable items included: newspapers, aluminum cans, cereal boxes, and more.

I spent a lot of time playing with this model. It's really quite fun. I must say I was surprised to find out how much energy was locked away inside my recycling. Knowing plastics are made from fossil fuels is one thing, but seeing it in real energy terms adds a whole new level of understanding. For instance, recycling ten 20-ounce plastic bottles, as opposed to throwing them in a landfill, saves enough energy to power my laptop for ninety-seven hours, or an incandescent lightbulb for eighty hours! Even better, a fluorescent lightbulb would brighten your life for over 370 hours!

The power locked in plastic bags is fun to think about, too. Ten plastic grocery bags will power my laptop for over three hours, and if we recycled every one of the estimated one trillion plastic bags produced globally each year, it would save over sixteen billion kilowatt hours (kWh) of electricity. For a point of reference, the average US home uses 11,280 kWh per year.

I wish all aspects of plastic recycling were as easy to understand and as fun as the iWARM model, but unfortunately, the recycling of plastics is often complex, inefficient, and rife with nuance and room for error. All sorts of questions arise when recycling plastic, some of which we've already covered, like, can I really ruin a whole batch of recycling by tossing the wrong number plastic into a bin? Or can I recycle anything made of plastic number 1, or is it just bottles? Can I toss the cap in? As stated in the text box in chapter 6, the golden rule with respect to plastics is when in doubt, keep it out! I know, I know, throwing stuff away when you aren't sure seems like the worst way to go about recycling plastic, but the fact of the matter is, it makes more sense in this case. Here's why. Mixing certain types of plastic recycling really can lead to contamination of the entire batch during processing.

Some mix-ups are quite likely. For example, PET (plastic number 1) is recycled all the time and PVC (plastic number 5) is rarely recycled, but they can appear very similar to each other visually and with respect to specific gravity. Specific gravity is the density of a substance relative to some reference substance, usually water at four degrees Celsius. If your specific gravity is less than one, it will float; more than one, it will sink. The similar specific gravity of PET and PVC makes it difficult to separate them mechanically, meaning they can easily wind up in the same batch of recycling because they float and sink together, even though they are made of completely different material and do not mix. PET and PVC are so incompatible that one PVC bottle in ten thousand PET bottles can ruin the whole batch. So, in short: *yes*, you can ruin a whole batch of recycling if you are careless, and that is why it's better to leave it out when in doubt.

Some other no-no's when it comes to recycling: no plastic bags. They can be recycled, but likely not as part of your single-stream recycling program. Single-stream recycling is a pretty great luxury if you are are lucky enough to live in a city or town that has it. It basically means you get to toss all of your recycling into one big bin with no need to separate the items by number or even by material (e.g., glass or plastic). Also, please don't go fill up a plastic bag with plastic bottles and place it in your bin. That plastic bag will mess things up pretty quickly by forcing workers to slow the conveyor belt, rip the bag open, sort the items, and start it going again. This wastes time and money. Also, no dirty diapers! No nonrecyclable plastics or Styrofoam! The point is, be careful!

Plastic number 5, Polypropylene (PP), comprises a lot of plastic packaging at grocery stores, mostly in deli items, like premade pizza dough, premade sandwiches, or small tubs of cheese. It turns out these number 5 plastic items aren't typically recycled in curbside recycling programs. However, there is some opportunity for recycling this kind of plastic at Whole Foods. Some of their stores have bins out to take back all that number 5 plastic. Which is a nice service, considering that, based on my experiences and observations, they seem to sell quite a bit of this less easily recycled plastic. And luckily, you can return plastic number 5 packaging from whatever grocery store you frequent at a Whole Foods recycling bin.

Remember, these tips are generalizations and guidelines, they aren't gospel. In the end, the answers to most of your recycling questions are specific to your town's or city's program. I've seen numerous articles stating that the plastic caps from containers like peanut butter jars cannot be placed into recycling bins. However, according to my local recycling rules it's a nonissue to leave those caps on for recycling.

While it may seem that opportunities for plastic recycling abound, at this point in time plastics recycling is still a terribly inefficient and incomplete operation with little to no opportunity existing for recycling certain types of plastic in many communities. Why is this?

Why is plastic recycling such a mess?
Can it get better?

Plastics have been referred to as "the last frontier of recycling." That's both exciting from an opportunistic standpoint and illustrative of the many shortcomings of the plastic recycling process in its current state. Even when plastics make it into the recycling stream, recovery rates are abysmally low relative to metals. The truth is, as a society we have a long way to go when it comes to plastics and recycling.

Recycling plastics is a complicated affair. An aluminum can is an aluminum can, any way you slice it. They might be different sizes, but they are all made from aluminum and therefore similar in density, melting point, and other physical aspects. The clear differences in magnetic properties and colors make automated mechanical separation of metals a snap. But we can't say the same for plastic. Not all plastic bottles, jugs, containers, and packaging are created equal. Depending upon the type of plastic used in their construction, they could have very different properties, yet appear very similar, each with their own density. Sometimes these densities can fall within a very small range to each other or even overlap. This makes separating plastics during the recycling process very difficult. Any plastic can also be any color, adding yet another complicating factor to the process of separation. Plus, plastic separation can't rely on differences in magnetic properties like metals can.

Despite these difficulties, different plastics need to be treated differently during the recycling process, adding considerable cost and complications, prohibiting the streamlining of recy-cling. Improperly sorted plastics, along with associated debris like paper labels or foodstuffs, can contaminate a batch of recycling and render the entire thing useless. The very unfortunate truth is that when it comes down to it, plastic is rarely recycled. And by rarely recycled, I mean, somewhere around 7 to 8 percent in the United States and 10 percent globally.

Although, as with most numbers and estimates for plastic production, consumption, and recycling, there seems to be a lot of uncertainty due to the sheer magnitude of what is being estimated. In addition to the approximately 7 to 8 percent of plastic recycled in the United States, about another 7 percent is incinerated along with other garbage to produce energy and heat. The rest of it, or roughly 27 million tons, gets sent to a landfill. Or worse.

There's an enormous amount of plastic that doesn't even make it to the landfill. Millions of tons of plastic instead end up in the ocean or littering our streets and beaches.

Plastic litter can persist in the environment for a really long time, up to a thousand years by some estimates. Of course, no one has ever observed a piece of plastic taking a thousand years to decompose, so that thousand-year value can be taken to mean basically forever in human terms. Plastic litter causes enormous problems with profound environmental and aesthetic repercussions. It's even been deemed responsible for human deaths. More about all that in chapter 11. But for now, let's explore the current and future state of plastic recycling a little more deeply, paying extra attention to the potential improvements and solutions.

As previously discussed, plastics are made from fossil fuels. When you see a plastic bottle floating down the river, you're quite literally looking at a solid hunk of oil floating downstream. It strikes me that we should be focusing on reusing that oil somehow. Seriously, why not recover the oil from plastics and help offset our need to drill for new sources of oil? This way we could also reduce litter and maybe even clean up some places that are already heavily contaminated with plastics.

Now I'm not saying we need to be thinking of new ways to burn more fossil fuels, but given the enormous amount of resources being sent to landfills and being littered, it isn't a stretch to contemplate a far more efficient vision while simultaneously working on expanding our renewable energy sources.

Plastics capture about half the carbon used to produce them, meaning they contain a lot of energy that could be used later. It isn't the easiest thing to fully grasp, but it does present some unique possibilities for using and reusing plastic several times and in several ways. It's complicated because we need to remember that reusing plastics to create energy doesn't free us from the fact that this is a product derived from fossil fuels, so we'll still be releasing carbon into the atmosphere. But if we used all that plastic for creating energy by, say, converting it back to oil, we'd be better off than having plastic waste and using resources to drill for more oil.

Apparently, I'm not the only one who thinks so, either. Columbia University prepared a report commissioned by the Plastics Division of the American Chemistry Council that aims to understand and quantify "ways to recover more of the energy value of non-recycled plastics (NRP) in the form of electricity, heat, or petrochemical feedstock. Landfilling of NRP constitutes a loss of a valuable energy resource. Capturing the energy value of non-recycled plastics will contribute to sustainable development and enhance national energy security." The key findings of the report are really interesting and can be summarized in one sentence: wow, there is a lot of wasted energy sitting in landfills!

Plastic Recycling—We've Got a Long Way to Go

1. Only about 10 percent of the plastic produced globally each year is recycled.
2. In the United States, 7 to 8 percent of plastic is recycled and another 7.5 percent is combusted and used for energy. So the rest, roughly 85 percent of plastic, ends up in landfills or as litter.
3. Depending upon the accounting, approximately 4.3 or 17.6 percent of the plastic bags produced each year are recycled. The 4.3 number is the 2010 EPA number for plastic number 2 bags. These are what you think of as

the typical grocery bag. The 17.6 percent represents number 2 and number 4 bags (e.g., bread bags, newspaper bags). (Apparently this is a touchy subject because misquoting this value has resulted in lawsuits by plastic manufacturers.)

But, is there enough wasted plastic to contribute to the enhancement of national security and sustainable development? I would think there would have to be quite a lot of energy locked away for that to be the case. Well, it turns out there really is. Columbia's report on NRP estimated that in 2008 the United States threw away 28.8 million tons of plastic; estimates for 2011 put this value around 27 million tons. We might as well just be pumping oil straight into landfills. Or we could skip the middleman and just landfill billions of dollars in cash, because 28.8 million tons of plastic is the energy equivalent of 783 billion cubic feet of natural gas or 139 million barrels of oil. That's enough to power all the cars in Los Angeles for a year.

The report also concludes that diverting 25 percent more waste from landfills to new, more efficient waste-to-energy facilities could reduce greenhouse gas emissions by thirty-five to seventy million tons of CO_2 per year. And this isn't just wishful thinking—the technologies really exist! There are public companies in existence beginning to do this very thing. One such company is JBI Inc. Their Web site states that company founder John Bordynuik first got the idea for the company in the mid-nineties when he began seeing an increase in the amount of plastic waste and decided a problem was coming. Their philosophy is an insightful one. The Web site offers this wise statement on the global plastic predicament: "Because of the new 'disposable mentality' that came with plastic containers, we now are navigating the fallout of an enormous waste plastic problem, on a global scale. It's complex, it's far-reaching and it's intimately tied to politics and economics." JBI's solution to the problem is Plastic2Oil technology. Using what they call "The Plastic-Eating Monster," JBI is essentially able to mine landfills for plastics and then process them back into common fuels used in industrial boilers and heating furnaces and oil-burning residential furnaces.

The New York Department of Environmental Protection finds JBI's technology to have close to a 90-percent recovery rate. The process does result in the production of some 10 to 20 percent as off-gases, but cleverly these are captured and used to power the process.

Another company out of Oregon, Agilyx, uses a similar patented, fully permitted process that converts ground-up waste plastic into synthetic crude oil, claiming that "an average of 8.5–10 pounds of plastic for one gallon of synthetic crude oil is a reasonable conversion factor."

As for resource production, both JBI and Agilyx seem to have created a pretty excellent process that wastes little energy and was designed with that specifically in mind. JBI keeps cooling water separate from the product to prevent contamination, using off-gases to power the process and electricity only to power the fans, pumps, and motors. In response to a question about carbon and energy footprints, Agilyx's Web site states, "Yes, the results show that the net carbon footprint of Agilyx's technology is favorable to traditional forms of crude oil extractions." Even though these processes still result in the burning of fossil fuels, JBI's and Agilyx's methods present an opportunity to tighten our resource consumption by eliminating the massive waste streams associated with throwaway plastic and reduce the risks associated with expanding the drilling and fracking efforts required to increase oil and natural gas recovery. Remember, I'm not claiming this is an entirely greenhouse-gas-free affair. It's not. Rather, it's a case of using oil twice: once to make plastic and once to burn for energy, just like regular old oil sans the plastic step. I'm also not claiming that these companies will succeed or stay viable in the long term, but they are fine examples of people pushing the envelope of energy technology.

The people at JBI and Agilyx aren't the only ones attempting to revolutionize modern recycling. Dr. Mike Biddle of MBA Polymers, an enormous plastic recycling company, gives a great TED Talk titled, "We Can Recycle Plastic." Biddle echoes the sentiments of JBI in stating that he is a garbageman because he hates waste. He highlights the severe inefficiencies in current plastic recycling practices, but fortunately he doesn't stop there, and why would he? People like John Bordynuik and Mike Biddle see waste as a resource and an opportunity. Why throw away all that energy that's locked up inside when instead they can contribute to cleaning up the environment and make money in the process? Biddle's company, MBA Polymers, has reengineered the way plastic is recycled, allowing them to create plastic for 80 to 90 percent less than the cost of virgin plastics while shrinking the size of waste piles, reducing greenhouse gas emissions, and providing human and environmental health benefits.

It's pretty much common sense when you stop and think about it for a second that plastic trash is cheaper than oil. The ability to take that trash and convert it into new plastic or even back to oil through an energy-feasible means (i.e., it doesn't require more energy to do it than you get in return) is a huge accomplishment and should be applauded as such. Other innovative companies are taking plastic waste and converting it directly into fun products. Literally. Safeplay Systems makes playgrounds from no less than 95 percent post-consumer HDPE plastic. Each playground they produce keeps 35,000-plus milk jugs out of landfills while providing kids with a fun place to play.

Companies like Agilyx, JBI Inc, Safeplay Systems, and MBA Polymers are representative of the clever innovation needed in this era of increased resource demand, declining resources, and heightened environmental pressures. Truly, there is so much room for improvement and really no excuse for not pursuing these oppurtunities.

Other countries are already recycling and converting energy from waste with remarkable efficiency. For example,

Japan is far better than we are at recycling plastic. They recycle a very respectable 77 percent of plastic consumed. Which makes our 7 to 8 percent percent look pretty shameful. Japan's recycled plastic is sent overseas to make toys and used in production of textiles, bottles, packaging, industrial parts, and a whole host of other products. Sweden also steps up the game when it comes to using waste as a fuel. In fact, they've become so efficient at converting waste into fuel that only 1 percent of their trash winds up in landfills. They even started running out of trash to convert, and began importing around 800,000 tons of garbage per year to create power and heat for homes.

Initiatives like these provide hope for the future because "recycle, baby, recycle" is smarter and sexier then "drill, baby, drill."

8.

BIOPLASTICS

Like the Kid Sister of Biofuels

Bioplastics (and biofuels) are becoming increasingly popular and easier to find in everyday life. In both cases, it's a seemingly wonderful and simple solution to a big problem, but when you dig deeper, in reality it's actually very complicated and not necessarily the panacea we may have hoped. This chapter deals with bioplastics, but there are so many similarities between bioplastics and biofuels that it may very well serve as a lesson on both.

Currently, bioplastics represent a very small (less than 1 percent) but rapidly growing part of the plastics market. Some estimates project growth as high as 30 percent per year (30 percent increase in the amount of bioplastics, not 30 percent of the entire plastics market).

Some big names in bioplastic manufacturing within the United States are NatureWorks—a collaboration between the

Japanese company Teijin and the US company Cargill—and Metabolix, based in Cambridge, Massachusetts. While still a very small part of the overall plastics market, bioplastics have benefited from both extremely high oil prices and the recent trend toward environmentalism and preference for "greener" products in consumer choices. The bioplastics market in the United States is estimated to reach a value of $680 million by 2016 and is expected to continue growing in the future. In 2012 both North America and Europe were using approximately 27 percent of their estimated available bioplastic production capacity, meaning companies in both countries have the ability to significantly step up production should they want to. Whether the bioplastics industry grows to fill those capacities remains to be seen.

Since bioplastics still comprise only an extremely small part of the plastics market (less than 1 percent, remember?), I think the most important question to ask is: do we really want that value to grow? And if it does, can it do so in a manner that makes sense from an environmental perspective? I know it may seem like bioplastics are ideal for sustainability, but the reality of the situation is that they aren't a cure-all for our plastic problem.

Have you ever heard the term "greenwashing"? Greenwashing is a form of spin in which marketing is deceptively used to make a company, policy, or product look environmentally friendly, when in reality it's anything but. Whether it is to increase profits or gain political backing, greenwashing's goal is to manipulate popular opinion to garner support for things or sell products that really aren't so green at all. I employ the term "greenwashing" here because both biofuels and bioplastics rely on it to a degree. Their marketing campaigns tend to boil a very complex issue down to a somewhat deceptive brand logo, like an image of a happy green cornstalk under puffy white clouds. I urge you to think back to the discussion of life cycle assessments in chapter 5 and remember to always consider the whole picture (or life cycle) when presented with seemingly green ideas, products, or politics.

With bioplastics, it's easy to leave out a large portion of the life cycle and come to some very erroneous conclusions regarding the costs. Often omitted is the amount of energy that goes into, and the negative effects related to, the growing, harvesting, and transportation of bioplastics. Also, in a world full of hungry people, do we really want to be pushing agricultural lands into production of plastics rather than food?

I'm not saying that bioplastics are all bad—they aren't. But realizing their shortcomings is critical to making informed decisions as consumers and citizens. To add another layer of complexity, just as with regular plastics, there are several types of bioplastics, each with their own characteristics to consider.

When I hear the term "bioplastics," I think it's some snazzy new invention, but these have actually been around for a really long time. Henry Ford even used corn-based plastics in the construction of the Model T in the early 1900s.

In modern society there are two popular types of bioplastic:

polylactic acid (PLA) and polyhydroxyalkanoate (PHA). Both PLA and PHA cost more than traditional plastics to produce. This is one of the greatest barriers to bioplastics expansion. For now, PLA costs about 20 percent more and PHA is nearly double the price of traditional petroleum-based plastics, but this could change. PLA and PHA are made using starches and sugars (e.g., cornstarch) and both degrade easily, or at least there are claims that both degrade easily. However, PHA can handle higher temperatures than PLA, making it more durable in a sense. Yet Metabolix, the primary manufacturer of PHA, claims that PHA is so biodegradable it will break down in streams and seawater without even being composted!

The issues surrounding the sustainability of bioplastics are, not surprisingly, quite complex. A particularly fascinating concern regarding bioplastics is of a behavioral nature. Steve Davies, the marketing director for NatureWorks, raises an interesting point about consumer behavior with regard to items like bioplastic bottles when he says, "They'll think, 'Oh, I can flip it out the car window and it will disappear on the side of the road,' and it absolutely won't." Even in the instances where people do pay attention and place bioplastic items in the recycling bin, they usually wind up in the landfill anyway.

Bioplastics get rejected from the recycling stream because the majority of recycling centers in the United States are still unprepared to handle them. Which means one of two things: either the bioplastic item ends up in a batch of regular recycling, where it becomes a contaminant, or it gets tossed into a landfill.

When bioplastics get tossed into a landfill they decompose in an anaerobic environment to form methane (CH_4), a greenhouse gas twenty-one times more effective at trapping heat than carbon dioxide. So you can see how that might not be a great thing. However, landfills can be equipped with the ability to capture methane and create energy to power all sorts of things. Many landfills in the United States are already equipped with this very technology. We actually excel in this department: "nearly 60 percent of the worldwide capture of methane occurs in the United States even though the US only generates 24 percent of the worldwide methane," according to a report on *The Importance of Landfill Gas Capture and Usage in the U.S.* by the Council for Sustainable Use of Resources at the Columbia Earth Engineering Center. The potential for this technology to mitigate greenhouse gas emissions and to generate power is quite astounding. The Intergovernmental Panel on Climate Change estimates "gas collection and utilization could reduce methane emissions from landfills globally by 70 percent at negative to low costs by 2030." The report by Columbia concludes using methane generated by landfills is a "win/win opportunity," the reason being that using the methane created in landfills to create energy converts that methane back to carbon dioxide, which is the very same gas that would have been created if the waste didn't end up in the landfill in the first place. And you can still obtain the primary energy benefit of the methane at the same time.

All this talk of landfills isn't to say that bioplastics are inherently unrecyclable, because they aren't. PLA is actually highly recyclable and can be recycled over and over again. But it cannot be mixed with the traditional plastic recycling stream. Therefore, a good method for separating bioplastics from petroleum-based plastics is needed. Germany, Switzerland, and Sweden require manufacturers to prominently mark products that are compostable so they can be identified and separated. The best separation method is infrared scanning, which can be used to differentiate between plastic types; unfortunately, this is an expensive technology that is out of reach for many of the small recycling facilities in the United States. Others have suggested a uniform color designation for compostable plastic products would help people to sort them.

Bioplastics don't necessarily free us from the concerns of endocrine disruptors either. For example, European Bioplastics states "its members are committed to avoiding the use of harmful substances in their products. Many plastic products do not use any plasticisers and a range of acceptable plasticisers is available if necessary. The wide range of bioplastics is based on thousands of different formulas. This means specific information regarding a certain material or product can only be obtained from the individual manufacturer, converter or brand owner using the material." So there is no uniform answer. They may have endocrine disruptors, they may not. Best to just avoid plastic all together.

Understanding the other potential downfalls associated with bioplastics should also inform your decisions.

A paper in *Environmental Science & Technology*, a well-regarded peer-reviewed journal, conducted life cycle analyses of several bioplastics versus traditional non-bioplastics and found some rather interesting results. The study rated the different plastics on a number of categories ranging from human health concerns, like potential to cause cancer, to environmental health effects, like global warming and eutrophication (the overloading of nutrients to waterways). This study used a "cradle to gate" assessment, meaning that it included the production of the plastics but not their use or disposal. In the case of assessing bioplastics, disposal may obviously have some benefits relative to traditional plastics. That is to say, what's the benefit of a plastic that composts rather than persists? Leaving out disposal seems kind of stupid, I know, but there is a good reason for this, which is that there is little quantitative information on the emissions and energy from degradation of bioplastics on which to base an LCA.

The assessment's results are a little bit surprising in that the bioplastics came out looking far less green than you might expect. The LCA conducted by *Environmental Science & Technology* found that bioplastics ranked most poorly in five of the ten categories assessed in this study, including ozone depletion, acidification, eutrophication, carcinogens, and ecotoxicity. That means that bioplastics, when assessed on a "cradle to gate" scale, may not always be so great, especially relative to more simple traditional plastics. The production of HDPE, LDPE, and PP did not result in the maximum impact in any category, likely because these are simple polymers that require less chemical processing than others. The authors make the point that "adhering to green design principles reduces environmental impact in either the petroleum or biological polymer categories. Switching from petroleum feedstocks to biofeedstocks does not necessarily reduce environmental impacts." The article also steps back and makes the recommendation that limiting bioplastics to those made from renewable resources that require little pesticide use or fertilizer might be an excellent thing for further mitigating the negative impacts bioplastic production. Harvesting invasive plants for such purposes comes to mind. I like the idea of cleaning up problems and producing plastics at the same time.

Bioplastics: The Solution to All Our Plastic Problems?

1. Right now bioplastics account for less than 1 percent of plastics.
2. Bioplastics aren't necessarily greener. Producing them may require more energy, cause more pollution, and detract from lands available for food production in a world with a lot of hungry people.

3. Advances and innovation in the production of bioplastics does make them promising. PHA plastics are produced by microbes and there are companies that claim to be producing carbon negative plastics without utilizing valuable cropland.

PART THREE

The Good, the Bad, and the Ugly Plastics

9.

THE GOOD

Benefits of Plastic

t's important to think about the big picture. It's also important to examine environmental and social problems within the context of reality and to search for feasible solutions. In researching this book I came across an article titled "10 Things We Wish Had Never Been Invented" and guess what number one was? Plastic.

Look, I will be the first one to admit that plastic causes a lot of problems on this planet and that our addiction to it is way out of control, but you will never catch me saying "I wish plastic had never been invented." Frankly, that's just silly. I'm not going to belabor this point, but I'd like to spend a little time reflecting upon the amazing role plastic plays in our lives and the ways it positively impacts every one of us every day.

Realizing the ways in which plastic benefits our lives can

occur on two levels: superficially and comprehensively. On the superficial side, it's easy to recognize some of the benefits of plastics. Plastics that make many, if not most technological advances possible, like the plastics that make up your cell phone or your computer, are obviously a huge benefit to our society.

Plastic quite literally saves lives, too. A lot of modern medical and scientific equipment and supplies are made from plastic, and while some of these items certainly have their shortcomings—for example, phthalates may be leaching from PVC IV bags into the bloodstreams of patients—those IV drips have still saved a whole lot of lives.

And it's not just IV bags that are manufactured from plastics. There are countless other items like plastic syringes, medical gloves, and other critical components of medical equipment. Not to mention two other big pieces of lifesaving plastic equipment: the defibrillator and MRI. Which brings me to my next point: science, as we currently know it, could not exist or continue to advance along these same lines without the assistance of plastic. Plastic *is* absolutely critical to our modern existence.

Plastic hasn't just facilitated advances in medicine and technology, it's also responsible for a wealth of environmental benefits. Yep, I said it: environmental benefits. I'm going to give credit where credit is due. Although these benefits may not be as easy to see as some of the negative images plastic might conjure up, like a seabird caught in a six-pack ring, the hidden environmental benefits of plastic do exist.

Many such benefits stem from plastic's high strength-to-weight ratio, which results in minimum packaging weights. Lower packaging weights translates to reduced transportation energy consumption. Additionally, the lightweight plastic materials used in vehicle construction further reduce the overall weight associated with transportation by reducing the weight of the vehicle itself. A lighter vehicle will require less fuel than a heavier one to travel the same distance. And it's been found that packaging beverages in PET versus glass or metal reduces energy use by 52 percent and greenhouse gas emissions by 55 percent during the transport of goods. Plastics also contribute to increased safety in transportation since they function as a central ingredient in airbags.

Beyond cars and trucks, the airplane industry also relies heavily on plastics to reduce the weight of planes and increase their fuel efficiency, which is essential given the already enormous carbon footprint associated with air travel.

Plastics contribute to numerous energy savings within the average modern home as well. Plastics are used in energy-efficient vinyl replacement windows and home insulation, both of which contribute to reduced energy costs and greenhouse gas emissions. They are also essential to the renewable energy industry. Because of their high abrasion resistance, self-lubricating properties, minimal thermal expansion, and good electrical insulation (among other things), plastics play a central role in green technologies such as photovoltaic cells—more commonly referred to as solar panels—wind turbines, hydroelectric power, and fuel cells. Solar panels, wind turbines, and solar water heaters can drastically reduce household energy consumption and/or accompanying emissions. These are just some existing examples, but there is enormous potential for plastic to continue making amazing contributions in the renewable energy sector.

As much as I will criticize the use of plastics in the modern food industry, I should also take some time to recognize the benefits that are imparted to human health and safety through plastic's ability to prevent contamination during shipping and processing. For instance, polyurethane foam insulates refrigerated trucks during shipping. I don't think anyone would want to eat a piece of chicken that traveled across the country in an unrefrigerated truck in the middle of summer (let alone food shipped from overseas). Even if you buy your meat directly from the people behind the deli counter and it appears to be sans plastic at the time of purchase, it was most likely packaged in plastic during transport to help prevent contamination and spoilage. And since food poisoning isn't that fun, we should be thankful for that kind of plastic packaging.

Truthfully, we are just scratching the surface in each of these categories when it comes to the benefits of plastic. The positive aspects and potential applications of plastic in transportation, technology, and medicine are pretty much infinite.

Plastic makes all sorts of great stuff even greater. From music to sex to skiing, plastic provides a means to create the products that can make these experiences more fun or even just plain possible. But, as we already know, plastic's versatility and affordability has a dark side as well—a dark side that manifests itself in excessive production and consumption of unnecessary, highly disposable, single-use plastic materials.

Unfortunately, at some point, we got lazy, lost our way, or were seduced by the convenience of plastic, and now we find ourselves on that plastic dark side. As a result, our use has spun way out of control. We use ridiculous amounts of plastic and create all sorts of waste in instances where it's completely unneeded. The consequences of this exorbitant usage are becoming disastrous for our health and the health of the environment we rely on to support society.

10.

THE BAD

Toxins and Plastic

If you don't have BPA in your body, you're
not living in the modern world.

—BRYAN WALSH, *TIME*

P lastic can be a human health hazard. Not all plastics,
but some. Depending upon the type or RISS number
of plastic you're dealing with, you can be exposed to a
range of toxins via different potential pathways. My hope in
this chapter is to familiarize you with the known toxins plastics contain and some of the pathways through which they
can enter your body so that the next time you go shopping you
can make some wise consumer choices when it comes to your
health.

It's infuriating and terrifying how many perfectly legal
products in our society are potential carcinogens or endocrine
disruptors. Some of them are more obvious than others. For
instance, it's rather easy to see why eating fast food three
meals a day might be bad for you or why smoking is a terrible
idea, but the health risks from plastics fly a little more under

the radar. Perhaps this is because we don't actively consume plastics. I mean, when was the last time you finished a bottle of water and then ate the bottle? The funny thing is, that is sort of what we're doing when we polish off a bottle of water! Plastics have the potential to leach toxins into your food and drink, and can even emit them into the air. Once in your food, drink, or air, they have an easy pathway for entering your body—a really scary thought. To help you keep those unwanted chemicals out of your body, here's some information about the toxins that are commonly found in plastic and the potential harm they can cause to your health.

Bisphenol A (BPA): BPA is an industrial chemical used in the production of hard clear plastics, like polycarbonate. Since 1960 it's also been used in the resins that line the interiors of food and drink containers. It was first made in 1891, but it wasn't used in plastics right away. Instead it was used as a hormone in cattle and poultry production and as an estrogen replacement in women! It is also a known endocrine disruptor.

Endocrine disruptors are defined by the National Institute of Environmental Health Sciences (NIEHS) as "chemicals that may interfere with the body's endocrine system and produce adverse developmental, reproductive, neurological, and immune effects in both humans and wildlife." The NIEHS goes on to say, "Endocrine disruptors may be found in many everyday products—including plastic bottles, metal food cans, detergents, flame retardants, food, toys, cosmetics, and pesticides." Which isn't good news, since all of us come into contact with at least some of these things on a daily basis!

The Food and Drug Administration's current perspective on BPA (as of the time of this writing) can be paraphrased as such: the FDA acknowledges that BPA is an endocrine disruptor and is present in foods and beverages associated with certain plastics but in quantities that are likely too small to be harmful; however there is uncertainty in this statement and more evidence is being collected. Should you want to read their stance in its entirety you can do so on their Web site.

It should be noted that the FDA did ban BPA from baby bottles and sippy cups, but only after the plastics industry pushed them to because of extreme pressure from consumers.

The FDA doesn't deny the fact that BPA makes it into our bodies. No one is really denying that. BPA metabolites are found in the urine of over 90 percent of adults tested in the United States. However, the FDA has concluded that the levels of exposure to BPA are too small to be worrisome, especially in adults. The EPA has also failed to express any major concerns regarding BPA exposure.

Interestingly, there is an enormous amount of peer-reviewed research suggesting that the FDA and EPA might be wrong in their blasé stance on BPA. But just what kind of damage might BPA do to you? It turns out that the effects are rather diverse but particularly scary for children because of developmental concerns.

Numerous and varied peer-reviewed studies have found evidence correlating BPA to "adverse developmental, reproductive, neurological, and immune effects" ranging from cancer to obesity. Of particular interest is the Chapel Hill Consensus, where thirty-eight of the world's top BPA scientists gathered in 2006 to discuss the scientific evidence surrounding BPA and possible deleterious health effects. And these aren't silly little side effects like a dry mouth. They ultimately issued a statement that said "BPA at concentrations found in the human body is associated with organizational changes in the prostate, breasts, testis, mammary glands, body size, brain structure and chemistry, and behavior of laboratory animals." The Chapel Hill Consensus also points out that over 90 percent of government-funded research had shown adverse effects from BPA, yet at the time, no industry studies had found any adverse effects. Hmmmmm . . . suspicious.

BPA and obesity have been linked in several studies. In one recent study, researchers from New York University examined the levels of BPA in approximately 2,800 children and young adults aged six to nineteen. The study aimed to determine if BPA levels in urine could be linked to obesity. The results from this study are pretty fascinating on several levels. First off, the study, which was published in *The Journal of the American Medical Association,* found that among Caucasian kids and teens, higher BPA levels were associated with a nearly doubled chance of being obese. Oddly enough though, in Hispanic and African-American children there was no correlation to BPA levels and obesity, which the researchers aren't able to make heads or tails of. I'd be remiss if I didn't mention that many other studies have published results linking BPA and obesity.

Frederica Perera, director of the Columbia Center for Children's Environmental Health, has said that "Obese children may be simply eating and drinking foods that have higher BPA levels." I think this quote speaks volumes about the role of plastic in our diet and how purging your life of plastic has health benefits on more than one level. There may be two things at play here: (1) BPA itself causes obesity and/or (2) foods with a lot of plastic packaging are generally not the healthiest options and tend to be more caloric, higher in fat, and generally worse for you. So from a consumer perspective, if you're concerned about your children's weight (and your own), then avoid food packaged with a lot of plastic!

As I pointed out earlier, obesity isn't the only negative effect BPA has been linked to. Peer-reviewed evidence has implicated it in all sorts of nasty things, including the nastiest nasty of all: cancer, especially, breast cancer. One study suggests that BPA in low doses may be linked to the onset of breast cancer, which is surprising and causes concern given the fact that the FDA and numerous other regulatory agencies have given statements saying that low concentrations of BPA are not dangerous.

In 2010 the United States President's Council on Cancer found that more than 130 studies had linked BPA and breast cancer (and potentially also prostate cancer).

And, as if cancer and obesity weren't bad enough, there's more. BPA has also been linked to negative effects on thyroid function, ADHD, and heart disease. These linkages aren't the easiest thing to establish. For example, since there are many causes and contributors to heart disease, like genetic predisposition, blood pressure, cholesterol, and weight, it's difficult to assess if BPA is truly a factor in the onset of heart disease. Nonetheless, the evidence is mounting.

Even though the evidence for a BPA–heart disease link is tricky, the connection between junk food, obesity, and heart disease are solid. Solid too, is the link between junk food and plastic packaging. Less plastic means eating less processed food, less BPA in your system, and a lower chance of heart disease.

When BPA isn't messing with your body, it's messing with your head. All sorts of crazy neurological effects have been associated with it. A 2008 Yale study found that BPA had caused the loss of connection between brain cells in primates at the EPA-safe level of fifty micrograms per day. The study summarized, "Our primate model indicates that BPA could negatively affect brain function in humans." The researchers concede that people would rarely be exposed to the daily doses of BPA applied in this study; however, they continue to note that BPA exposure is cumulative over a lifetime and therefore effects may be cumulative, too.

With all this negative evidence linked to BPA, it does seem somewhat strange that the FDA, EPA, and others are so casual about the potential negative effects of it on humans. Other countries have acknowledged its dangers. Canada has even gone so far as to label BPA a toxic substance, while France has passed a law banning it in food-contact packaging effective in 2015, positions I wish the US government would take.

New York Times reporter Nicholas Kristof suggests that the chemical manufacturing industry is following in the footsteps of Big Tobacco by preventing regulation of their industry despite evidence of adverse health effects for millions. Kristof hopes, however, that new studies that suggest transgenerational effects from very low doses of BPA may change this in the future. The particular study that he referenced in his column was published in the *Journal of Endocrinology* and found that low-level doses of BPA administered to mice resulted in less sociable mice being born. The researchers even found that the mice exhibited signs that parallel those of autism in humans! And while the United States regulatory agencies may be slow to act, there is no reason why you can't protect yourself and your loved ones on your own. Why risk it?

Common BPA Exposure Pathways

Some common exposure pathways are via:

- ♻ Airline boarding passes
- ♻ Canned foods
- ♻ Canned soda and beer
- ♻ Microwaving foods packaged in or stored in plastic. Heating foods in plastic increases the chances of endocrine disruptors like BPA (and phthalates) making their way into your food. If you are going to microwave food, I recommend doing it in ceramic or Pyrex.
- ♻ Non-BPA-free water bottles and even "BPA-free" water bottles
- ♻ Receipts

Phthalates: Phthalates are oil-derived plastic esters, often called plasticizers, which are used to increase the flexibility, transparency, durability, and longevity of plastics. Phthalates have been in use as a plasticizer for roughly fifty years and are the most commonly used plasticizer in the world, with about two million tons being produced annually. More than 75 percent of the US population has phthalate metabolites in their urine.

The EPA has phthalates on a list of chemicals that "may present a risk." As far as plastics go, phthalates are what allow plastics like PVC, or plastic number 3, to be so soft and flexible (think of a shower curtain). Like BPA, phthalates are also classified as endocrine disruptors and have been implicated in numerous human health concerns—most recently, diabetes in women. And as with BPA, they also seem to have the most profound effects on children, particularly young boys. Phthalates are thought to be responsible for emasculating infant boys and have been linked to abnormalities including increases in hypospadias (a birth defect where the urethra forms incorrectly) and undescended testes.

Much like BPA, research has also proposed links between phthalate exposure levels and breast cancer, which is terrifying. Especially when considering that phthalate levels tend to be higher in women because of their presence in so many cosmetic products.

Congress has actually banned the usage of certain phthalates in children's toys after a study conducted by doctors at the University of Rochester Medical School found a correlation between phthalate levels in the bloodstream of pregnant women and the health of baby boys they gave birth to. The study demonstrated that as phthalate levels increased in women's blood, so did the health problems of their baby boys. The regulations enacted by Congress permanently banned any amount of three types of phthalates (DEHP, DBP, BBP) in children's toys and certain child care products.

While phthalates have been banned from children's products in the United States, the American Chemistry Council has, unfortunately, deemed them to be totally safe despite all the evidence to the contrary. Nonetheless, some companies have begun to respond to consumers' concerns, and it seems that Avon, Johnson & Johnson, and SC Johnson have plans to remove phthalates from their products under their own initiatives. The state of California has also banned products for children and babies that contain more than residual quantities of phthalates.

Avoiding phthalates is actually incredibly difficult, as they are in a lot of products and exhibit that sneaky plastic spirit of appearing in the most unexpected places. But hey, that doesn't mean we can't safeguard ourselves. Although often used as a plasticizer, phthalates appear in nonplastic items as well. Due to that we're going to step outside the range of plastics and talk briefly about a few other products here as well. Phthalates are used to make PVC (aka polyvinyl chloride, or plastic number 3) flexible, which seems straightforward enough since you can easily identify the difference between soft and hard plastics. Soft plastic number 3 is a likely place for phthalates. For example, even though I applauded the fact that medical equipment is made from plastic, a disturbing fact is that hospital IV bags and the tubing for those IV bags contain phthalates. In situations where premature babies have been administered IVs, the phthalate levels in their blood have spiked enormously. Given that these endocrine disruptors seem to affect infants and children most dramatically, you would think that we could come up with a better alternative.

But consider this: Phthalates are also used to hold fragrances to products and to help make lotions more spreadable. And, oddly, they are used as a coating on medicine pills and vitamins. Does anyone want to eat some phthalates and then slather themselves in them?

My initial thought upon learning about the widespread use of phthalates was that fragrances weren't too tricky to avoid, especially since I don't wear cologne or perfume. But then I realized that fragrance included things such as scents in detergents and soaps. Moreover, cosmetics are loaded with this stuff. Yikes. Phthalate concentrations in cosmetics might be responsible for the generally higher levels of phthalates found in women relative to men. The use of phthalates in cosmetics is not regulated, but the manufacturer or company must include them on the ingredients list on the label.

You might be thinking, "Great, I can easily avoid phthalates now. I just need to read the ingredients lists on labels." Oh man, how I wish it were that easy, but alas it's not. Unfortunately, if the phthalates are part of the fragrance as opposed to some other part of the product, there is this obnoxious loophole in the law that allows the phthalates to be left off the ingredients list. This makes it difficult to know for sure whether phthalates are in the product you're using. See, your soap might not contain phthalates directly, and therefore it doesn't need to say so on the bottle, even though the fragrance used in the soap, and therefore the soap itself, contains them. This makes avoiding phthalates incredibly difficult. Best bet: Avoid scented soaps and lotions.

Some Unexpected Places That Phthalates Might Be Hiding

- ♻ Vinyl raincoats
- ♻ Shower curtains
- ♻ Rubber duckies
- ♻ Car dashboards
- ♻ Steering wheels
- ♻ Car deodorizers
- ♻ Lotions
- ♻ Detergents
- ♻ Lipstick
- ♻ Nail polish
- ♻ Plastic sex toys
- ♻ Hairspray
- ♻ Paints
- ♻ Medicines
- ♻ Vitamins
- ♻ Furniture
- ♻ Food packaging
- ♻ Air fresheners
- ♻ Insect repellent
- ♻ Flooring
- ♻ Rocket propellant—yeah, bummer, I know. So try to cut back on your consumption as much as possible.

Note: Not all products in all these categories will contain phthalates, but I cannot possibly cover each instance. Now that you are aware of the potential you can read up on your own prior to making a purchase. Good luck!

In general, understanding phthalates and the risks they pose is a little more complicated than understanding BPA. This is because there are numerous types of phthalates that have different codes assigned to them. Much like the numbers used to identify different plastics, knowledge of these codes can go a long way to keeping you safe and informed.

Some of the Common Phthalates

- ♻ DBP and DEP. These are commonly found in toiletries and cosmetics, like nail polishes, deodorants, perfumes and colognes, shampoos.
- ♻ DEHP. Used in PVC plastics and some medical devices.
- ♻ BzB. Used in flooring, car parts, and some toiletries.
- ♻ DMP. Used in insect repellent and soft plastics.

Styrene and Benzene—A Whole Other Reason to Hate Polystyrene. Styrene is a precursor to polystyrene (plastic number 6), which we most commonly think of and refer to as Styrofoam. Technically, Styrofoam is a trademarked extruded polystyrene product created by Dow Chemical Company and used largely as insulation. People, however, commonly use the term Styrofoam to describe any number of expanded polystyrene products, such as coffee cups or takeout containers.

The classic reason for hating Styrofoam is because it takes a ridiculously long time to decompose—estimates range from one hundred to four hundred to one million years (point being, it's not going anywhere fast)! And since most Styrofoam can't be recycled, a lot of it ends up in landfills (or littering our environment) and stays there for between one hundred and one million years. Expanded polystyrene, or what we think of as Styrofoam, can't be easily recycled because it's an end product. That means it can't be reused for something else, as it's too labor-intensive and toxic for most recycling centers to handle the process.

About 6.8 million tons of styrene is produced globally each year. Styrene is a "hazardous chemical" that causes problems upon contact with the eyes or skin or upon ingestion or inhalation. The EPA has said that it's a suspected toxin to the gastrointestinal tract, kidneys, respiratory system, and more. And if that isn't enough, the US National Toxicology Program has said that styrene is "reasonably anticipated to be a human carcinogen"—although no regulatory agencies have as yet officially declared it to be a carcinogen.

This matters much because styrene can leach from polystyrene plastics into your food and drink. Polystyrene can also leach benzene, another really nasty compound. The Department of Human Health and Services has determined that long-term exposure to benzene can cause cancer. It also causes weakening of the immune system, decreased red blood cells, anemia, and possible adverse developmental effects on babies. Animal studies have identified all sorts of crazy effects related to benzene exposure, including low birth weights and bone marrow damage.

11.

THE UGLY

The Environmental Costs of Our Plastic Addiction

How much brighter and cleaner a world [it would be]
than that which preceded this plastic age.
— V. E. YARSLEY. AND E. G. COUZENS, *PLASTICS*
PENGUIN BOOKS LIMITED, 1945

When thinking about or discussing environmental issues, I think it's important to remember that they are not separate from social and economic issues. Our culture relies directly on the ecosystem services provided by the environment to sustain our lives and the economy. Therefore, environmental degradation often has direct economic and human health costs associated with it. Environmental degradation caused by plastics is no exception. While some economic and human health impacts are quantifiable (i.e., they can be transferred into numbers such as monetary values or number of illnesses), sometimes less quantifiable things like aesthetics are important, too.

It's tough to assign a cost to the knowledge that sea turtles exist, right? As equally tough as it is to assign a cost to beautiful fall foliage, but we all know that these things have value in some way. The aesthetics of a nice clean living space has value, as do clean beaches, rivers, and parks. How many of you have been on a walk and seen a plastic bag in a tree, or half buried in mud along the river, or have driven down a road with a bunch of gross plastic litter along the shoulder? All of you, I'm sure. We all see these types of things all the time and it's easy to become desensitized to them, but that doesn't mean that we should stop trying to prevent them. Many of us don't accept these things in our own individual yards, so collectively we must make an effort to not accept them anywhere. And it's important to remember that the damage from plastic waste is far more than just aesthetic. It leads to very real and very serious problems.

While it's easy enough to see plastic litter in the streets, parks, or even wildlands of the United States, the epidemic becomes far more apparent in developing nations. Traveling in Cambodia, Egypt and Peru, I was blown away by the amount of plastic waste that littered nearly every patch of land I set foot on. I fully recognize that not every country has or can afford to have a well-developed or even any public-waste infrastructure, and as a result the amount of litter that piles up can be truly astonishing.

In developing cities and nations, the lack of sanitation services means most plastic bag waste ends up in the street, and all those bags choke storm canals, cover drains, and clog pipes, so water doesn't get where it needs to go and can cause floods. I've seen it myself in Cambodia, where heavy rains turned the streets into a knee-deep plastic soup in a matter of minutes.

This poses problems on a number of levels, some of which have very destructive consequences. The amount of plastic waste became so severe in Mumbai, India, in 2005 that it choked the city's drainage system and resulted in severe flooding that caused the deaths of over a thousand people. Plastic bags have been implicated in several major flooding events in Bangladesh during the late 1980s as well. By banning bags, these municipalities hope to prevent such future flooding, or at least mitigate it.

In Cape Town, South Africa, plastic bags have been banned, and some extremely harsh laws have been passed that would impose heavy fines and even prison terms for vendors caught distributing plastic bags to customers. The law aimed to reduce bag consumption by 50 percent in a nation where they use about eight billion bags per year, many of which end up floating in lakes, flapping in trees, and stuck in fences. Their plastic bag litter had gotten so out of control that they jokingly referred to them as "the national flower."

Flooding and unsightly aesthetics aren't the only negative side effects associated with unchecked plastic bag consumption and a lack of sanitation infrastructure. In fact, when plastic bags clog drains, they can alter local hydrology and create microclimates conducive to mosquito breeding, leading to increased spread of mosquito-transmittable disease, including malaria and dengue fever, endangering human populations.

Another indirect cost of plastic litter is its negative effect on tourism and its detriment to local and even national economies as the rubbish piles higher and higher, making the landscape unsightly. These costs are tremendously difficult to quantify but they are no doubt a reality.

Because plastics are so inexpensive, lightweight, and durable they are widely used in packaging and single-use disposable products. In fact, 50 percent of plastic is used in the production of single-use plastic items. The prevalence of plastics in human society, combined with their highly disposable nature, makes for a situation where a lot of plastic items never make into the proper waste stream and end up in the world's oceans and on its beaches. It is estimated that plastic makes up an estimated 60 to 80 percent of marine debris. Entanglement and ingestion of plastics may lead to death—especially for sea turtles, who tend to mistake plastic bags for jellyfish and try to eat them. In fact, some research suggests that ingestion of plastic debris may be the leading cause of sea turtle mortality.

And it's not just turtles that eat plastic bags and get sick. In the year 2000, three thousand cattle mysteriously died in Lucknow, India, and when the city investigated, it was revealed that the culprit was plastic bags. The cattle had been ingesting plastic bags, which clogged their stomachs and resulted in a painful death. Remember that cows are sacred in India, so this was a big deal. Other nations like Kenya and Somalia have reported numerous cattle deaths due to plastic bag consumption, which is hard on the livelihood of those people.

Since approximately 49 percent of plastics are buoyant, and so many of our major cities are along rivers or the coasts, when plastics are improperly disposed of or escape from poorly maintained landfills, they can be easily transported via waterways to the ocean. Not only are plastics transportable, they are persistent. Interestingly, since plastics have only really been used in mass production since the middle of the twentieth century, no one really knows how long they are capable of sticking around in the environment. But we do know the rate at which they enter the environment is far, far greater than the rate at which they break down. Therefore, they accumulate over time.

The amount of plastic entering the ocean on an annual basis is mind-blowing and truly difficult to conceive of, but I'll give you the numbers anyway. Six point four million tons of plastic litter make their way into the ocean each year! This is an estimate that the National Oceanic and Atmospheric Administration (NOAA) came up with in the 1970s prior to a moratorium on ocean dumping, but still.

A commonly cited but somewhat dubious value also says there are 46,000 pieces of plastic debris floating in the ocean per square mile. More reliable data is available for the amount of plastic in the ocean gyres, and the numbers are an order of magnitude higher. Yikes. If you aren't familiar with the phrase "order of magnitude," it just means you tack another zero on the end. So we're talking hundreds of thousands of pieces of plastic per square mile covering an area estimated to be some-where between twice the size of Texas and twice the size of the continental United States. Quite a range, I know, but es-timating the size of a shifting, vast, and remote area of plastic in the middle of the ocean is extremely difficult.

The gyres themselves are large areas of open water where currents converge and rotate to draw in debris, including floating plastic waste. There are five gyres, but the most well-known one is the North Pacific Gyre, which has earned a sad reputation as the Great Pacific Garbage Patch. This isn't to say the other gyres don't contain stunning amounts of plastic pollution, because they do.

The name Great Pacific Garbage Patch, coupled with the knowledge that, on average, hundreds of thousands of pieces of plastic are present per square kilometer (a mile is 1.6 kilo-meters) makes it easy to conjure up an image of a huge tangled island of garbage bobbing along on the ocean surface. In fact, such images have even circulated in the media, but that's not the reality. Despite all that waste in the North Pacific Gyre, the garbage patch isn't visible from space or even the air. The reason for that is that the garbage patch is basically a soup comprised of suspended plastic pieces "about the size of your pinkie fingernail," although bigger items are certainly present in the mix.

But where does all this plastic come from, and why are the pieces all so small? Good questions. While researching this section, I ran into an oft-cited but difficult-to-confirm statis-tic, stating that 80 percent of marine debris comes from land

and 20 percent from marine sources. Again, NOAA points out that while this is most definitely possible, it's difficult to track the origins of these values down. Regardless of the ratio, it's estimated that eight million items of marine debris enter the oceans every day through varying sources. Once in the ocean, plastics do break down, but not necessarily in the way you'd think. Rather than dissolving, they fragment into smaller and smaller pieces over time, until one day you end up with oceans that have turned into a plastic soup and sea life that is constantly consuming that plastic. And then you may end up consuming that sea life!

Plastic waste in the oceans has even been implicated in the death of whales. A sperm whale in Spain was found dead on a beach after consuming 59 plastic items weighing a total of 37.4 pounds. Much of this plastic was in the form of thick plastic sheeting that had clogged the whale's stomach and resulted in its untimely death. Plastic litter killing an animal that weighed 4.5 tons. Insane.

Fresh water ecosystems suffer from the effects of plastic waste too. The Great Lakes are currently being inundated with the tiny plastic microbeads contained in beauty products such as facial scrubs, body washes, and toothpaste. These beads help scrub away dead skin, or plaque when contained in toothpaste. They are designed to be small enough to wash right down the drain, making it really easy for them to wind up floating in the Great Lakes. Researchers found up to 600,000 pellets per square kilometer in Lake Erie. Fish readily mistake these pellets for food and ingest them, which has to make you wonder about eating fish that eat plastic. As one researcher said, "We don't know if the problem stops with the fish or if when we eat the fish, the problems are with us now."

By now you may be fairly freaked out by the potential for plastic to adversely affect your own health, the health of your children, and the environment at large. If you're feeling overwhelmed and helpless about all the plastic in your life after all my blabbing about plastic being everywhere and unavoidable, don't be! Let me show you why. . . .

PART FOUR

Time to Purge Some Ugly Plastic!

12

GETTING STARTED

Plastic is here to stay, and as I said earlier, in some cases that's not a bad thing. However, the amount of useless and unnecessary plastic we consume needs to decline dramatically. Luckily, because such a staggering amount of plastic is used in nonessential packaging and single-use applications, the opportunities for cutting plastic waste are easy and plenty. Not only is dropping these plastics fairly simple, but cutting back on them may greatly improve the quality of our lives on personal and societal scales.

Let's make something clear here. I'm not going to tell you to never purchase anything with plastic packaging again. If you can swing that, it would be pretty amazing and I'd consider you a total rock star! But my purpose within this section is to provide you with a few simple rules, tips, and tidbits of knowledge to help you become more aware and more capable of an easy and successful plastic purge. Whether that means swearing off all single-use plastics forever or simply getting a few reusable grocery bags, that choice is up to you. So take what you learn here and come up with a mix that works for your lifestyle, but remember, the more the better.

So here's how we are going to go about it. Throughout the remainder of this book, I'll present numerous tips to help you cut back on plastic. Not all of these will be equal in their ease of adoption or their impact on your health and the environment. Some of the easiest things we can do when it comes to plastic have the potential to make a really big difference. To help you get a quick sense of the amount of effort required to purge some plastic, you will find between one and three little plastic bottle symbols next to each section. The number of bottles corresponds to the level of effort required to make the change and purge some plastic.

One bottle: This is an action that everyone can do with a minimal effort. I'd go so far as to say that this is stuff you *need* to be doing. It's about removing the most unnecessary plastic from your life and the environment.

In some cases, this will help lessen your exposure to toxic chemicals, too.

Two bottles: 🍶🍶 These actions and tips require a little time and effort, but are well worth it. At this level are some of the actions that I think will help you lose weight and feel healthier. If you give a little more, you get more benefits. Depending upon your shopping habits, some of these actions might be a big change, or you might find that they're small, worthwhile sacrifices.

Three bottles: 🍶🍶🍶 This level is for people who want to experiment with cutting out as much plastic as possible, as I did during my experiment for Grist. Trying out the actions at this level, even for a little while, will make you much more aware of plastic in your everyday life and will also help you create less waste.

So there you have it, some simple classifications to help you understand your options at a glance. Throughout the following chapters I will do my best to show you how to make these changes. It would be great if I could point out every product and every option on the market and rate them all for you, but that's really not possible given the incredible number of items available. I'm also not sure I want to provide such a level of detail within this book. There are instances where I have mentioned specific products, but products can change or better ones can become available in the market. Therefore, I'd rather teach you the kinds of products to look for and general tricks for avoiding plastic so that you can keep your life as plastic free as possible far into the future, and I can have a book that's more relevant than just a catalog of specific items. Now let's get to work.

13.

GROCERY SHOPPING

Here's an interesting little tidbit: Eating fresh foods, such as those from the farmers' market or even just the produce section of the grocery, can cut the amount of BPA in your body by 50 to 70 percent after three short days. In chapter 10, we touched on some studies that made the link between BPA and obesity, suggesting that fetal exposure to BPA may preprogram our bodies for weight gain and obesity later in life. We've also covered BPA's links to cancer and reproductive health issues. In the next few paragraphs I'd like to talk about the less science-y and more commonsense link between plastic packaging and weight.

Here's the radical but obvious idea I'm putting forth in this book: Junk food uses more plastic packaging, while real, healthy food uses less. Let's look at Kraft Singles as an example. Do we really we need to wrap every slice of cheese in an individual plastic package? Think about it for a second. Would you rather have fifty individually wrapped pieces of "pasteurized prepared cheese product" (What does that even mean?) or one piece of plastic wrapped around a whole wedge or wheel of actual cheese? Or more specifically, something the FDA allows to be labeled as straight-up cheese, not "cheese product." I'm not just picking on Kraft Singles, because when it comes down to it, pretty much all highly processed junk foods are packaged in plastic. Just picture all those bags of potato chips, Hostess snacks, cookies, sugary cereals, candies, cake mixes, and sodas. A lot of this stuff lives mainly in the middle of the grocery store, which is a processed food and plastic packaging wasteland.

Given the nature of the modern food industry, it's nearly impossible to avoid plastic at the grocery store. When foods and goods are being processed and transported all over the world, there's going to be packaging. And plastic is the packaging that tends to keep things from spoiling for the longest amount of time. But fortunately, there are some pretty easy ways to drastically cut your plastic consumption and limit your exposure to nasty chemicals. It really just takes some know-how and a little bit of effort.

Before I delve into a discussion of the grocery store, I want to point out that the best way to avoid single-use plastics and the not-so-savory toxins that may accompany them is to avoid the modern grocery store altogether. Realistically speaking, I know that isn't always possible. Nonetheless, I'd like to say a little something about getting food from outside the grocery store from places like farmers' markets and even wholesale clubs. (Oh, and if you're looking for information on baby foods and products, you'll want to skip ahead to chapter 16, or if you're looking for information on beverages like beer, water, and wine, read on in this chapter.)

Farmers' Markets 🍼🍼: A great way to begin purging the plastic from your life is by visiting your local farmers' market as frequently as possible. Not only does the farmers' market offer the best chance of purchasing foods with minimal packaging (and minimal processing), but you can actually meet the people who grow what you eat. And trust me, strolling through the farmers' market on a Saturday or Sunday morning is a hell of a lot more pleasant than walking around a grocery store.

In addition to fruits and vegetables, many farmers' markets also sell bread, cheese, and meats. Not only does the food from the market have less packaging, but because it's local it tends to be really fresh and far healthier, and I guarantee you it will taste a million times better than what you get at the grocery store.

If you don't know where the closest farmers' market is in your location, you can visit http://search.ams.usda.gov/farmers markets. This is a really handy site maintained by the United States Department of Agriculture that allows you to search for farmers' markets by zip code.

When and if you do find that farmers' market by your house, you can bring your own bags or a box to toss your food in. And there you have it: delicious plastic-free food. It's that simple. Oh, and don't forget to bring cash, because most markets aren't going to be able to run your credit card, although that seems to be changing with the advent of phone and tablet compatible credit card processors.

Wholesale Clubs 🍼🍼: While in many ways the idea of wholesale clubs like Sam's Club or Costco may seem the opposite of green shopping, since there's so much stuff packed in those stores, they actually offer some advantages that are worth considering. As Jeff Yeager proposes on the Web site thedailygreen.com, these stores shouldn't be written off so quickly, especially when it comes to plastic and packaging. First off, they force you to bring your own bags or boxes to carry groceries in, rather than giving you free, disposable ones. Buying in bulk can also mean fewer items individually wrapped in plastic if you shop smart.

However, while the farmers' markets and wholesale clubs may provide some great options for cutting plastic consumption, eventually we're all going to wind up at the grocery store. Below, I've tried to give some tips for navigating your grocery store from the perspective of reducing your plastic consumption.

The Modern Grocery Store

Ever heard the saying "shop the edges"? Shop the edges means avoiding the middle of the grocery store, because it is mainly full of junk food, which as I pointed out above, is usually packaged in plastic. The edges are usually comprised of the produce, bakery, deli, and butcher sections of the grocery. These areas all tend to have products with a lot less packaging than the center of the store, and also contain foods that tend to be less processed. While I was doing my plastic purge experiment, it was this connection that resulted in Mary and me losing weight and feeling really healthy during our purge. And who isn't psyched to shrink their footprint and their waistline?

So you've got the big picture, right? Stick to the edges of the store and beware the junk food in the middle. This is going to be our guiding principle as we navigate the modern grocery store and cut the crappy foods and plastic from our carts. I'm not saying that you can never eat junk food, but rather, I'm proposing that you might find that awareness of your plastic consumption can give you a new lens through which to think about your shopping cart.

Now it's time for some specifics.

Reusable Grocery Bags 🍶: Let's start really basic: Avoid the evils of plastic bags (see chapter 11). Go buy some reusable grocery bags. You can get them at most grocery stores or order them online. Actually, you can get them just about anywhere these days. We have around ten of them in our home, but four or five seems to do the trick for a week's worth of groceries for two people. Personally, I recommend buying a thinner nylon bag with longer handles, especially if you plan on walking to the store from time to time. The longer handles make placing a full bag of groceries over your shoulder on the walk home from the store more comfortable.

Since we're talking about bags, it might occur to you to ask the question: What about just using paper bags instead of plastic, because trees are a renewable resource and paper is biodegradable and can be recycled? Well, actually, the EPA found that the production of paper bags requires more energy and results in more pollution than the production of plastic bags.

I came across some articles positing that the plastic grocery bag was created to help ease the pressures of paper bag production on forest resources. Which it may have, but clearly there are a host of other problems that have arisen from them.

Surprisingly, the plastic bag wasn't really used widely until the 1980s, and only came into existence in 1977. That means when I was born (1976), there was no paper versus plastic option at the grocery store. We'd be better off if things stayed that way and the plastic bag hadn't been invented, because today it's estimated that globally one hundred billion to one trillion plastic bags are used annually. That's a lot of plastic!

And making plastic bags isn't a cost-free endeavor by any stretch. As you know by now, plastic bags are typically made from HDPE or plastic number 2. HDPE is in turn made from fossil fuels. In fact, the energy in about nine plastic bags is enough to drive a car about three quarters of a mile.

Also, plastic bags have remarkably low recycling rates. The EPA estimates roughly 5 percent of number 2 plastic bags in the United States are recycled annually, while globally the number may plummet to around 1 percent. (Interestingly, despite seeing this 1 percent value cited over and over, I cannot determine the original source.) That means that here in the United States some 95 percent of number 2 grocery bags wind up in the landfill or worse: blowing or floating around or stuck in a tree or along the side of a river or lake. The California Coastal Commission estimated during a cleanup of the Los Angeles River that plastic bags and films represented 43 percent of all plastic collected.

Other cities all over the world have already addressed the unnecessary consumption of single-use plastic bags, whether it be through outright bans or bag fees. However, here in the United States, there's been strong opposition to such bans and bag fees by retailers and the plastics industry. As a result, the discussion around plastic bags has focused mainly on taking the bags back and recycling them, which unfortunately hasn't been working too great. But there is hope! Cities around the United States are beginning to join the ranks of other countries all over the world, such as Ireland, Germany, Switzerland, and even China, who recently put a nationwide ban on plastic bags.

Evidence shows that these bans really work. For instance, Ireland's bag tax reduced their plastic bag consumption by 90 percent. And in an era where reducing dependence on foreign oil or oil altogether is a good idea, an action like this can gain traction. According to Chinese estimates, the first year of their bag ban saved 1.6 million tons of oil. So there's a huge incentive.

The beauty of the plastic bag fee versus the outright ban lies in its ability to also address retailers' concerns about potential costs as they adjust their cash register stations and accounting systems. By charging ourselves a small price at the register (for those times when we screw up and leave our bags at home), we can retrain ourselves to remember our reusable bags, all the while helping to fund the minor adjustments required by merchants, and also contribute toward the purchase of reusable bags for those struggling to make ends meet. Bag fees can also be expanded to include paper bags as well, which, as I said earlier, actually have a bigger footprint than the plastic ones, at least during production.

In a *Wall Street Journal* article titled "Should Cities Ban Plastic Bags?" one of the arguments made against banning plastic bags centers around a report from the UK Centre for the Environment that found a reusable cotton grocery bag would have to be reused 173 times in order to match the energy saving of one plastic bag (due to the costs associated with production and transportation of cotton). However, in typical lie-by-omission fashion the writer neglected to mention the fact that the report also states that nonwoven PP (polypropylene) bags need to be reused only four times. It also fails to mention the obvious fact that we use far more than one plastic bag per year or trip to the grocery store. More along the lines of 1,500 plastic bags per year per family! Plus, you can fit a lot more groceries in a reusable bag than you can in a disposable one. When you think of it that way, making up for those reusable bags doesn't take too long even if it's cotton.

Another negative that is sometimes voiced in association with reusable grocery bags is their potential for becoming hotbeds of microbes and disease. Studies have found both clean and dirty bags. Surprise, surprise. If this is something you are really worried about (I'm not at all), then just give your bags a wash from time to time. Problem solved.

Plastic Bags—A Big Problem and a Total Waste

1. Between one hundred billion and one trillion plastic bags are used globally each year—so many we can't keep track.

2. Plastic bag recycling rates are abysmal, with the EPA estimating something like 4.3 percent of number 2 plastic grocery bags were recycled in 2010. When they are recycled, plastic bags can cause serious problems for recycling facilities, costing millions of dollars per year by clogging recycling equipment and forcing plants to shut down while the clogged machinery is fixed.

3. One to two million plastic bags are consumed globally each minute, a staggering volume.

4. Cutting out plastic bags is one of the easiest ways to start reducing your plastic footprint.

Disposable shopping bags really are the ugliest of ugly plastic. If you do nothing else that I recommend in this book, you should at least switch to reusable bags. It's so easy! And if everyone did it we could eliminate so much waste with nearly zero effort. Plus, you'll always have an extra tote kicking around for your next beach trip.

The Produce Section: Let's start our shopping in the produce section because it's generally the first section one encounters when entering the grocery store. Plus, it happens to be along the edge of the store, which as we mentioned earlier, tends to be where less plastic and more healthier foods are found. That doesn't mean the produce section is entirely plastic free, but relative to the rest of the modern grocery store, it's pretty darn close.

I'll tell you what tripped me up the first time I hit the produce section with the intent of fully avoiding plastic. I went to grab some greens beans and then I walked directly over to the plastic bag dispenser in the produce section and said, "Crap crap crap!" I can't believe I didn't think of it until I was standing there like an idiot with a handful of green beans. But that's the thing—our plastic use has become so second nature that we don't even notice it until we take a step back and really try to kick the habit. We're like a bunch of plastic junkies.

No worries, though, because there are some really easy ways around this. Which one you choose depends on how far you want to take your plastic purge.

Reusable Cloth Produce Bags 🍶🍶🍶: If you're feeling super motivated and want to completely eliminate plastic bags when you go grocery shopping, you can purchase reusable produce bags. They're just light cloth bags the size of the plastic ones we all rip off the spool in the produce section, but you can use them over and over again and toss them in the washing machine when they need to be cleaned. A five-pack of these will cost you between ten dollars and fifteen dollars and be more than enough to handle your produce needs. Today many grocery stores sell them right in the produce section.

The first thing you need to learn about putting your produce in cloth bags is that you'll need to know how much a bag weighs beforehand when it's completely empty. Otherwise, two ounces of lettuce might end up costing you the price of six ounces of lettuce. There are two ways to go about getting the weight of your bag; both are super simple and take about five seconds.

1. Look at the label when you buy your bags because it might just tell you.
2. Weigh one using the scale in the produce section. I just took a rough estimate of the average weight and then told the cashier the amount when he weighed the bag with my produce in it. Then the cashier can subtract the weight from my purchase.

Skip the Produce Bags Completely 🍶: A great way to cut back on bags is to simply not use them at all. Unless you are buying small, loose items, like mushrooms, you don't have to use any produce bags. Just toss that pepper or those apples right into your cart or basket.

Make the Most of the Produce Bags from the Store 🍶: For many of us, remembering to bring reusable grocery bags to the store in the first place is enough work, and adding a whole other set of bags that require weighing is more than we have time for. Well, that's OK! I completely hear you. Fortunately, there are still some really easy ways to cut back the plastic and save yourself some money without doing any additional work, or needing to remember anything. All you need to do is reuse those plastic produce bags at home.

Prior to my Grist experiment, before I'd even heard of using cloth produce bags, we'd gone years just reusing our plastic produce bags, empty bread bags, and other things like empty sour cream containers to meet our plastic bag needs. It really works and it's really easy. Either way, you'll be using less plastic and spending less because you won't be purchasing those Ziploc bags to store your food (which, by the way, count as ugly plastic) if you haven't yet switched to reusable food storage containers.

Keeping Produce Fresh Without Plastic Bags

Note that dry cloth bags cause leafy produce like lettuce to wilt, so here are some tricks for keeping your lettuce and other veggies happy without storing them in plastic bags:

1. Moisten your bags slightly with water and leave the lettuce in them. This isn't my favorite method because it's a bit messy and keeps you from being able to use your bags whenever you want to, since they'll be occupied, but if you are in a rush this method works well.
2. Use a salad spinner. When you get home, simply chop that lettuce up, spin it, and put in the fridge inside the spinner to keep it fresh. However, if your salad spinner is freaking huge and takes up the entire refrigerator, then see point 3.
3. Take your lettuce and place it in a suitably sized bowl. Then cover it with a moist paper towel, or better yet, a moist cloth napkin to keep things from wilting.

4. Mason jars are also great for storing leftover chopped-up vegetables.

5. Stainless steel or Pyrex bowls with lids are wonderful at preserving produce and come in all sorts of shapes and sizes to accommodate everything from huge salads to half an onion.

The Bulk Section: This is a fantastic way to cut plastic packaging from your grocery cart. What's available in this section of the grocery store varies depending on where you live and where you shop, but the general idea remains the same no matter where you are. I recommend the bulk section for getting everything from spices and flour to granola, pastas, beans, and rice. The food available in bulk sections also tends to be reasonably priced, as you're not paying for packaging or fancy labels. And an added bonus: It's all pretty healthy stuff. I've never once seen a bulk bin of Twinkies.

Bring Your Own Containers 🍶 🍶 🍶: To make the bulk section completely plastic (and paper) free, you can bring your containers or bags to the store with you and fill them up. But don't forget that, as with the reusable cloth bags, you'll want to know how much your bags/containers weigh beforehand or you may end up overpaying.

Use the Bags Provided 🍶: Even if you just use the bags that your grocery store provides, you can still make a big dent in your plastic consumption by shopping in the bulk section. Just reuse that bag at home and you'll be another step closer to never buying another Ziploc bag ever again.

Canned Foods and Hidden Plastics: Canned foods sure are handy. Just about everything comes in a can: soups, beans, sauces. Which is why it is such a bummer that canned food is a fantastic entry point for BPA to get into your bloodstream. Yuck! But all hope isn't lost, because you've got options and I'm going to lay them out there for you.

Avoid Metal Cans Altogether 🍶🍶: Nearly anything you can get in a metal can you can get in a glass jar or simply make using fresh ingredients. For example, to-mato sauce is always also available in a glass jar. Green beans—use fresh ones.

Switch to Tetra Pak but Beware 🍼: You know those cardboard containers that you sometimes see soup or things like soymilk or almond milk packaged in? They are Tetra Pak. Tetra Pak doesn't leach nasty BPAs into your food or drink, but it does have other problems. The main one being that it's difficult to recycle because it's composed of numerous layers of cardboard and foil. As a matter of fact, it's not even an option in *most* communities in the United States. The manufacturers of Tetra Pak claim that they are working to improve recycling opportunities in the future, and hopefully they will. For the time being, check with your community's recycling program.

Buy from Companies Who Have Elected Not to Use BPA in Their Can Linings 🍼🍼🍼: Some items like soups are very convenient in can form. And while it's great that you can find cans with BPA-free liners from companies like Eden Foods, Muir Glen, and Trader Joe's, the choices are still few and far between. Plus, these more select brands' products aren't always available or affordable. During the writing of this book, Campbell's Soup announced that it would be going BPA free as well, but as yet they have not announced what they are going to replace the lining with or when exactly the replacement will be completed. Eden Foods uses a "custom-made can lined with an oleoresinous c-enamel that does not contain the endocrine disruptor BPA. Oleoresin is a mixture of oil and resin extracted from plants such as pine or balsam fir." Eden Foods points out that while their BPA-free lining works great for products like beans, there still isn't a great alternative when it comes to acidic products like tomatoes, other than glass.

Unfortunately, some of the alternatives to BPA lining might be equally as bad. There is a helpful Word-Press blog at http://bpafreecannedfood.wordpress.com /bpa-free-canned-food-brands/ that lists the most up-to-date information. My best advice is to avoid the cans altogether.

The Dairy Section: The three places to cut down on plastic in the dairy section are when you're buying milk, yogurt, and cheese. Let's begin with milk. Unfortunately, milk containers are tricky when it comes to purging plastic. You actually have quite a few options here: cardboard, plastic, glass (sometimes). So which one should you purchase? Glass is a great option, but one that can eliminate itself in a lot of instances because it's not always available and tends to be really expensive. But hey, if you can find it and you're willing to spend the extra money, then by all means, go for it!

Weirdly, the glass jugs are truly the only non-plastic option when it comes to milk, because even those cardboard half-gallon milk containers are lined on the inside and out with plastic. So if you can't score your milk in a glass container, then I actually recommend purchasing your milk in the plastic gallon containers. I bet you didn't expect to hear that! But seriously, it seems to make more sense for a couple of reasons. First off, most plastic milk containers are made from HDPE plastic, which means they won't leach nasty chemicals into you. Secondly, plastic milk containers are easily recyclable in most communities and their manufacture and transport requires less energy than the cardboard containers. Plus, the cardboard containers are really just plastic anyway.

Ironically, most of the organic milk comes in the less environment-friendly cardboard containers, but it is possible to find plastic gallons of the organic stuff at Whole Foods and other groceries. I personally think the idea of milk from a cow that's all hopped up on antibiotics and growth hormones is disgusting, so I'll take my organic milk however I can get it.

Opt for the Big Tubs of Yogurt 🍶: Yogurt is another interesting product that seems to have some lessons to teach us. I've yet to see a yogurt that comes in anything but a plastic container, which raises an interesting question. Should I buy it at all? Well, that's up to you, but I think it's important to recognize that even when plastic is the only option there are still ways to do a little purging. For starters, consider buying a larger quart-size yogurt and spooning individual servings into a Pyrex storage container to bring along to work or pack in your kid's lunches. Also, double-check to see that the container you are purchasing is recyclable where you live. Stonyfield Farm now makes some plant-based yogurt containers that are also recyclable and nonleaching. You can also reuse larger yogurt containers in place of plastic bags to store food before you ultimately recycle them when the lids begin to wear out.

Frozen Foods: I hear you, frozen food is really convenient. The trick here is to stick to the guidelines we've already learned. Microwavable meals are wrapped in an extra layer of plastic that I don't recommend nuking, as heating plastics can increase the chances of endocrine disruptors making their way into your foods. Plus, these meals also tend to be high in sodium and bad fat relative to freshly prepared ones. You might also want to avoid products that contain lots of individual wrapping, like say, individually wrapped frozen snacks.

Opt for Fresh Rather than Frozen 🍼: Bags of frozen veggies are easily avoided if you just buy fresh. It may require a little more prep time on your end, but you will be repaid with a better-tasting meal. Otherwise, just be mindful of the plastic you're putting into your cart and what it's saying about your overall health.

Beer and Wine: What kind of crappy beer and wine comes in plastic bottles? you may be wondering. Not very many are bottled in plastic (although I think I once had a Miller Lite in a plastic bottle at a baseball game), but beer and wine is another place where hidden plastic and potential health dangers come into play. And no, I'm not just talking about the little six-pack ring. We'll start by considering the plastic in your beer, as it is far more common encountering it in beer than in wine.

Go for Glass Beer Bottles 🍾: I love a can of cheap beer as much as the next guy or gal. In fact, we are a country that is "in the midst of a canned beer renaissance" according to Consumerist.com. And unfortunately, nearly all those aluminum beer cans are lined with an epoxy resin that contains BPA, which results in fantastic headlines like this one from TreeHugger.com: "More Americans Drinking BPA in Canned Beer, Thanks to Economy and Pabst Drinking Hipsters."

Joking aside though, there is BPA hiding in aluminum beer cans, and that is nothing to laugh about. But how much BPA is in there, and should you really be concerned? A study by Health Canada examined seventy-two different canned beverages (not beer, but aluminum cans nonetheless) and found BPA in 96 percent of the cans tested. However, concentrations were extremely low. So low that consuming enough BPA to meet the personal daily intake limit set by Health Canada through this avenue alone would require consuming 940 canned beverages in a day. On one hand, if you're consuming 940 beers per day, then BPA is the least of your worries. On the other hand, there are still a lot of unknowns when it comes to what level of BPA exposure is actually safe. (For more information see chapter 10.) If you decide to play it really cautious, then by all means drink beer from bottles until a better and safer liner is available in the future. On the upside, both glass bottles and aluminum cans are easily recycled just about anywhere.

Cut Up Plastic Six-Pack Rings Before Recycling Them: How can we discuss beer and plastic and not mention the infamous plastic six-pack ring? I don't think it would be a stretch to say that for a lot of us the first image that comes to mind when we think about problems resulting from plastic is a bird all tangled up in a plastic six-pack ring. This problem has received a lot of attention at various points in the past, resulting in legislation being passed in 1989 requiring that all plastic six-pack rings be designed to photodegrade after about ninety days. That's certainly better than nothing, but I personally wouldn't want to spend a month or three tangled up in a plastic sixer ring. Plus, when improperly disposed of, those six-pack rings break down and form all those tiny little bits of plastic that spread throughout the oceans winding up in fish, animals, and even our food (see chapter 11).

The best thing you can do with your plastic six-pack ring is cut the loops and then recycle it. Traditional six-pack rings are made from LDPE (plastic number 4) and can be handled by many local recycling programs, but as always, be sure to check before you recycle.

Stick with Wine Bottles Rather than Wine in a Box 🍶**:** There isn't a whole lot to say about wine and plastic, but there are a few things. When wine and plastic do meet, it's in the case of wine in a box. Inside the outer cardboard layer is typically a polycarbonate (plastic number 7) bag capable of leaching BPA into your wine. So if BPA remains a concern of yours, you might want to stick with the bottle. It's also possible that your wine was aged in casks lined with BPA-containing materials. This is trickier to discern and may require some digging prior to purchase.

14.

ELIMINATING PLASTIC AT HOME

n the kitchen: Plastic is everywhere when it comes to storage within the kitchen. From the small plastic sandwich bag to the small Tupperware for leftovers to big gallon bags for freezing things. If there's one thing that will cut your plastic consumption when it comes to food storage in your home, it's mason jars.

Use Mason Jars 🍶: Mason jars have been around since the 1800s and are still available for purchase today. That's because they rock. If you're not familiar with mason jars, you might not realize that they come in all different sizes, which can serve a whole slew of different functions. There's also something quite stylish about them. They are rather pleasant to look at and to hold, much nicer than a plastic container in both look and feel.

We purchase our mason jars at the local hardware store and they are very reasonably priced (around a couple bucks apiece depending upon the size). We buy them in four different sizes and each one serves a different purpose. Mason jars also go excellently with bulk purchasing, which is another great way to eliminate plastic waste. Here's the breakdown on how we use the different sizes in our household:

Two-quart mason jars—These are the biggest ones we have. We use them for storing things like rices, bulk pastas, coffee beans, and flour that we purchase in the bulk section. Not only does it cut packaging waste, but we don't need plastic storage containers. An added bonus: They line up nicely along the backsplash of our kitchen counter, and the different items contained in them lend a ton of color and texture to the room. It's really quite pretty.

One-quart mason jars—We use these similarly to the two-quart version, but for things we purchase less of, such as bulk teas or nuts to snack on.

Twelve-ounce widemouth mason jars—These serve a dual purpose in our house. Firstly, they are our drinking glasses and they work wonderfully as such. They are far cheaper than buying a set of glasses and they have a whole lot more functionality. Since the jars comes with caps, you can also use them to mix up and store homemade salad dressings. They are also great for transporting smoothies or snacks with you to work for later in the day.

Two-ounce size—These cute little jars are perfect for storing spices from the bulk section. They even come with labels for you to write on to help keep your spices in order. These also look really nice in the cupboard with all the different-colored spices. You can also use them to bring a little salsa or salad dressing with you in your lunch.

Mason jars are no doubt central to cutting plastic consumption in the kitchen, but because of their shape, there are some things mason jars aren't great for, like storing a leftover piece of chicken or leftover pasta. It could work, but it's not ideal. Plus, you might not want to jam your kid's sandwich into a round mason jar. Anyway, this is where Pyrex comes in.

Buy Some Pyrex Containers : I love Pyrex for storage. It handily takes the place of both plastic wrap and Tupperware storage items. Pyrex storage containers are surprisingly reasonably priced and come in all shapes and sizes, from small enough to take with you in your lunch to large enough to store a big head of lettuce or leftovers from an enormous meal. Plus, they have tight-fitting lids to help keep things fresh or from spilling during transport. The lids themselves are plastic, but considering how much you can reuse them, this small amount of plastic doesn't bother me much. Also, Pyrex claims that their lids are BPA free. I should also mention that you can swap the Pyrex containers for stainless steel and everything applies. Stainless steel might be better in some instances, as you don't have to worry about containers shattering.

Since the sizes and shapes of these containers vary, they're really handy for storing different items. The rectangular Pyrex containers are perfect for blocks of cheese, leftover pizza, carrots, sandwiches, and whatever else you might previously have put in a Ziploc bag, while the more bowl-shaped ones will do the trick for soups, pastas, and salads.

Between mason jars and Pyrex you should eliminate the need for plastic bags, plastic storage containers, and plastic wrap in your kitchen. As an added bonus you might even end up beautifying things a bit when your bulk goods are displayed in lovely glass jars on the counter.

Get Some Stainless Steel Ice Cube Trays : Most ice cube trays I see are plastic. However, I feel like I commonly see lots of fancy refrigerators that are smart and industrious enough to just make ice cubes for you.

One place you will not find a fancy ice-cube-producing refrigerator is in our kitchen. There were some ice cube trays in our fridge when we moved in. However, since we're not certain whether they are BPA free or not, we won't be freezing our baby's food in those nondescript ice cube trays. There are some BPA-free plastic ice cube trays and stainless steel trays that can easily be ordered online. However, since BPA-free doesn't necessarily translate to safe (as discussed previously), I would recommend avoiding plastic altogether whenever possible when it comes to ice cubes.

Stop Buying Bottled Water 🍼**:** This section comes with a caveat: Bottled water is an incredibly complex topic. Tap versus bottled water regulation alone is worthy of an entire chapter in a book. And since entire books already exist on the subject of bottled water, my goal here is to give you the basics and the important highlights.

Much like the single-use plastic grocery bag, bottled water represents the ugly side of plastics. Admittedly, bottled water is handy during big emergencies, like natural disasters or storms that knock out power and otherwise prevent delivery of potable water. But for the most part, we as a society don't drink it due to an emergency. We drink it because we're lazy. One survey found that 56 percent of U.S. consumers cited convenience as the number one reason they drink bottled water. Once you see bottled water for what it really is, purchasing it becomes tough to follow through with.

Bottled water is convenient. But bottled water wasn't available until 1990. So I *know* we can exist without it in modern society.

Unless you're traveling internationally and can't drink the local water, or in the midst of a crisis and it's the only thing available, there's really no reason to buy the stuff. And a note about international bottled water: It's not always safe. Depending upon the country, the quality of bottled water can be terrible! I recommend referring to your travel guide to get a sense of what to drink where.

Right now you may be asking, why would bottled water not be clean? What about in America? Isn't bottled water always safe to drink? Isn't it cleaner than tap water? Not necessarily. Bottled water is in some ways less regulated than tap water. Many of these regulatory differences are due to the fact that bottled water isn't classified as water. Rather, it's regulated by the FDA as a food. These differences were significant enough to propel the Government Accountability Office to publish a concisely titled report by the name of "FDA Safety and Consumer Protections Are Often Less Stringent than Comparable EPA Protections for Tap Water." The title of the report explains it quite well, but let's examine some of the finer discrepancies between tap and bottled water in detail.

The Safe Drinking Water Act provides the EPA with the authority to mandate testing of public tap water by certified laboratories and requires public water entities to report on the quality of the water to their customers. Both of which they do. On the other hand, bottled water suppliers are not required to disclose the source of their water. They're also not required to report on the quality of the water to their customers.

Beyond these obvious failings are other less obvious but equally egregious ones. Many of which originate from a big loophole in the classification of bottled water as a food rather than as water. Since bottled water is a food, it is subject to FDA regulations, but FDA regulations apply only to food products sold in "interstate commerce." This loophole may exclude 60 to 70 percent of bottled water! Regulation of the bottled water exempted by this loophole falls under the purview of individual states, if they have any. Paraphrasing a report from the Natural Resource Defense Council:

> 13 states told NRDC that they have *no* resources, staff, or budgetary allotments specifically earmarked to implement the state bottled water programs and twenty-six states reported having *less than one* full-time person dedicated to running the state's bottled water program.
> —Statement from Frank Shank, director, FDA Center for Food Safety and Applied Nutrition, reprinted in "Bottled Water Regulation," at a hearing of the Subcommittee on Oversight and Investigations of the House Committee on Energy and Commerce, Serial No. 102-36, 102nd Cong., 1st Sess. 65, p. 75 (April 10, 1991). (www.nrdc.org/water/drinking/bw/chap4.asp #note117)

As if these shortcomings aren't enough, the FDA doesn't sufficiently regulate the amount of coliform bacteria in bottled water. Coliform bacteria, including the deadly *Escherichia coli*, is found in human and animal waste. In recognition of this, the EPA has strict rules that require frequent testing for coliform on tap water supplies, up to 420 times per month in public water sources serving over 2.5 million customers. Smaller public water supplies are tested on the order of 60 times per month. The EPA uses total coliform tests as the first layer of safety. If total coliforms are detected, then subsequent testing is employed to determine if they are of a dangerous nature. If they are dangerous, then customers must be notified the same day and the water must be treated appropriately.

The FDA rules for bottled water are not as stringent. FDA standards call for testing only once per week and permit the allowance of a small number of total coliforms without confirming the coliforms present are benign. There are no rules requiring bottled water companies to report contamination or recall the product.

To be fair here, the International Bottled Water Association (IBWA) does have strict safety standards and encourages its members to comply with their code and accept annual inspections. Which, as the NRDC suggests, is commendable, but also flawed in that adhering to the code is voluntary.

If regulation (or a lack of) isn't enough to turn you off, there are plenty more reasons to dislike bottled water. First off, it's just silly. Think about it for a second: A company takes something you can essentially get for free, puts it in a single-use plastic bottle, adds a label to it, and sells it back to you at an enormous markup. It's one of the greatest scams of all time. The EPA estimates that if you drink your recommended eight glasses of water a day from the tap it will cost you a whopping fifty cents per year, whereas if you bought solely bottled water for your hydration needs, it would be $1,400 annually. By this accounting, bottled water is roughly 2,800 times more expensive than tap water! Other estimates I've seen place the price of bottled water around 1,000 times more than tap water. And the craziest part is, there's a great chance that you'll be buying tap water in a bottle anyway! Between 25 percent and 47.8 percent of bottled water is actually just tap water, depending on the estimate.

Bottled water also has a ton of other not-so-fantastic characteristics associated with it. Categorically, they revolve around the much greater energy and resource demands associated with bottled versus tap water. *The Onion* hits the nail on the head with the headline of a fake news story they published called, "'How Bad for the Environment Can Throwing Away One Plastic Bottle Be?' 30 Million People Wonder." We drink a lot of bottled water and throw away a lot of bottles. The actual number of bottles of water consumed annually by Americans is way higher than thirty million. In 2011, US consumption of bottled water peaked (so far), hitting 9.1 billion gallons.

The majority of bottled water (think one-liter size) is sold in bottles made from plastic number 1—polyethylene terephthalate, or PET for short. PET is relatively safe and the leaching of nasty chemicals from PET isn't a concern. The big five-gallon water cooler jugs, on the other hand, are not so safe. These tend to be made from plastic number 7. Plastic number 7 is a bit of a catch-all category, but in the case of five-gallon water jugs, we're typically talking about polycarbonate plastics. Polycarbonate plastics, including those five-gallon home and office water coolers, have the potential to leach BPA.

As for the resources associated with our bottled water

addiction, it mainly uses up both fuel and water. You would think that the only water being used in a bottle of water is the amount being held in the bottle, right? However, for any volume of bottled water produced, twice as much water is used in its production. So every time you buy a liter of water, you're effectively wasting two liters. As you now know, plastics are made from fossil fuels. Thus, in accounting for the cost of producing bottled water, we need to include the energy costs associated with these fuels. Producing the bottles and associated packaging for all that bottled water is estimated to use seventeen million barrels of oil annually, not including transportation energy costs, which have been estimated to be as high as filling each one liter bottle of water with oil.

What about recycling? Recycling surely helps alleviate the resource consumption involved in producing bottled water. Well, of course it does, and you can read more about that in chapter 7. However, even if you recycled every bottle of water you purchased throughout your lifetime (realistically this number is closer to 12 percent of all plastic bottles produced in the United States), you'd still be wasting two liters of water for each one you purchased, and you'd still be spending at least a thousand times more for a potentially crappier (literally) and less safe product than you can get out of the tap.

Just as another example of how wasteful and costly bottled water is at the time of writing, a liter of bottled water in the United States is up to three times more expensive than a liter of gas. Yet people rarely complain about the price of water.

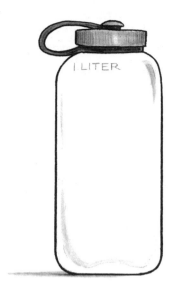

The Reusable Water Bottle : Thankfully, there's a really simple and convenient solution to all of this. Buy a little water bottle and carry it with you. You can fill that reusable bottle repeatedly, save yourself a whole lot of money, and if you buy an aluminum or stainless steel water bottle you can also spare yourself all those endocrine disruptors. To be on the super-safe side, I'd go for a stainless steel water bottle, as even the aluminum ones sometimes contain mystery linings. Note that I didn't say to use a BPA-free reusable plastic bottle—not because such a thing doesn't exist; nearly any store that sells reusable water bottles will sell a reusable BPA-free version. One small problem, however, is that BPA-free plastics may be just as bad, or even worse, than the products they aim to protect us from. Bisphenol-S (BPS) is one of the replacements for BPA, and unfortunately, BPS has also been shown to have some of the same concerns as BPA. The same is true for BPB and BPF.

So next time you consider buying that bottle of water with the label showing snowcapped peaks, remember there's a good chance it's tap water, and if it's not tap water, then it's likely less regulated than tap water, and it's costing you thousands of times more!

Bottled Water Is for Suckers

1. Bottled water is less stringently regulated than tap water.
2. Between 25 and 48 percent of bottled water is tap water.
3. Bottled water is 1,000 to 2,800 times more expensive than tap water.
4. It's estimated that about seventeen million barrels of oil, not including transportation energy costs, are used annually in bottled water production.
5. Bottled water requires 2,000 times more energy than tap water to produce the same amount.

Drink Soda from Glass Bottles or Opt for the Two-Liter 🍾: Soda cans contain some BPA, whereas plastic soda bottles, like the 20-ounce or 2-liter kind, don't contain BPA, so I guess if you need your soda fix you may want to consider drinking it from 2-liter bottles (which are highly recyclable!)

The Reusable Coffee Cup 🖋 🖋: A *Scientific American* article points out that "It takes 77 million years to make fossil fuels and 45 minutes to use as a coffee cup."

I'm a total coffee fiend. When people tell me they are giving up coffee, I feel almost personally insulted. But I admit that disposable coffee cups create a lot of waste. Each year Americans throw away two hundred billion disposable cups! Many of these were surely used to hold our precious coffee. Around twenty-five billion of those cups are Styrofoam, yet another possible avenue by which chemicals like styrene can make it into your body.

Even non-Styrofoam coffee cups contain a thin plastic lining that probably goes unnoticed by many of us. And that plastic lining causes big problems when it comes to recycling. Starbucks alone sells around three billion disposable coffee cups a year, and nearly all of them wind up in landfills. To their credit, Starbucks did lobby the FDA to get these cups to be 10 percent post-consumer recycled material, and has pledged to make 100 percent of their cups recyclable by 2015, so things should be on the up-and-up for them. But given the fact that most other materials that can be recycled aren't, I'm not sure how helpful this will be. Not only do all these disposable paper cups lined with plastic create waste, but they may also leak BPA into your body through that poly-ethylene liner.

Despite this bad news and my absolute love of coffee, there's really no reason to fret, because the solution is incredibly simple: Buy a reusable coffee cup or buy a couple! Keep one at home and one in your office, or in your work bag or car.

Reusable coffee cups are so easy to use and so fantastic. Why fantastic? They are fantastic because not only do you eliminate a bunch of waste and lower your exposure to nasty chemicals, but your coffee stays hot way longer! And the deliciousness of coffee, as we all know, is positively correlated with its temperature. Not only do you win on that front, but a good reusable coffee cup will never have a leaky drippy lid like the disposable ones sometimes do. Plus, as a final bonus and nod of approval for your efforts, you get a discount on coffee when you bring your own cup, so you will eventually pay for your new disposable.

I recommend the Starbucks stainless steel ones. They do a great job of keeping your coffee hot and they don't leak or drip. And the best part is, there's no plastic waste.

Reusable Coffee Cups—A Must for the Eco-Conscious Coffee Fiend! 🍶

1. Disposable coffee cups contain plastic. Nope, that's not wax.

> **2.** Hundreds of billions of disposable cups are thrown away each year in the United States. Many of them are used for coffee.
>
> **3.** Reusable coffee cups are super easy and beneficial on so many levels. By purchasing a reusable coffee cup, you can have hotter, cheaper, and less wasteful coffee. Reusable coffee cups also don't leak or spill as easily as disposable ones, which is great news if not spilling on yourself is as much of a struggle for you as it is for me.

Other Storage in the Home:

Use Baskets and Metal Bins 🍶: Storage outside the kitchen is less of an issue. We do keep some of our camping gear in plastic bins, but only because it's the best option. If the bin gets dirty we can just wipe it down, plus it's durable and cheap (two of plastic's defining characteristics). But one thing I really dislike about plastic is the fact that it looks cheap. While it serves its purpose well enough tucked in the closet and packed full of water filters, crampons, and camp stoves, I never really want to look at it in my house. Some other nicer-looking plastic-free options are baskets or galvanized steel washtubs, which are excellent for storage and are very cool-looking. We chose these over plastic, in many cases. You can also use them to do your laundry, and they make fine hampers.

15

PLASTIC AND PERSONAL HYGIENE

A long with food, most of us also make regular purchases of personal hygiene products such as soaps, shampoos, deodorants, toothbrushes, and toothpaste. If you take a quick look around your bathroom, I think you will find that most of these are packaged in or made from plastic. So what options exist to purge some plastic from your personal hygiene products and not wind up smelling really bad? Quite a few, actually.

Shower Curtains: Who doesn't enjoy a shower after a day of working or playing outside? I know I do! I also enjoy a clean bathroom and a new shower curtain from time to time. But, hey, you have to wonder about that new shower curtain smell. . . . You know, the really plasticky one that comes pouring out the second you tear open the packaging. Is that smell toxic?

The answer is, most likely very much so! I say "most likely" because if you have a polyvinyl (PVC, plastic number 3) shower curtain, which many are, then you have a pretty damn toxic item in your bathroom. That smell, the one that comes pouring out of the package when you open it, is a plume of toxic gases that have built up while the item was sitting on the shelf. An EPA report on PVC shower curtains found that they could result in elevated concentrations of air toxins for over a month! Similarly, a 2008 report by the Center for Health, Environment & Justice titled *Volatile Vinyl—The New Shower Curtain's Chemical Smell* found that PVC shower curtains resulted in elevated concentrations of toxins in the air for twenty-eight days and reached levels of toxicity sixteen times higher than the guidelines for indoor air quality established by the U. S. Green Building Council. Incredibly, that study found PVC shower curtains released over a hundred volatile organic compounds (VOCs) into the air, including all the same ones found in the EPA assessment. Of the 108 chemicals released by the curtains, seven are classified by the EPA as hazardous air pollutants. The report points out that VOCs can cause irritation to your eyes, nose, and throat, headaches, nausea, and even damage to your central nervous system, liver, and kidneys. A shower curtain from Walmart contained VOC concentrations of twenty thousand parts per billion, so high the analytical instrument was maxed out and analysis had to be halted.

The report also found PVC shower curtains to contain high levels of the phthalates DEHP and DINP. Recall that phthalates are no good at all and have been linked to impaired reproductive development in males, sperm damage, and premature breast development in girls along with a whole host of other things (see chapter 10).

As if this isn't all scary enough, these experiments did not replicate the humidity and temperature conditions present in a bathroom during a shower. The authors concluded that under those conditions the air concentrations would likely have been higher.

Sufficiently freaked out yet? You should be. Can you do anything about it? Of course you can! Avoiding all these toxins is as easy as knowing which shower curtain to buy (or not to buy) because you can get safer ones just about anywhere. To purchase a safer shower curtain is pretty simple, and a number of choices abound depending on whether you want a safer plastic or no plastic at all.

Get Some Cloth Curtains 🍾🍾: Should you want to forgo the plastic all together, you can opt for a cotton shower curtain. Or even an organic cotton shower curtain. Unlike plastic shower curtains, cotton ones take on moisture and will need to dry out so they don't mildew. As a result they may work best in dry rather than humid climates. You may be thinking to yourself, won't a cotton shower curtain result in my bathroom being a sopping wet mess when the cotton fails to stop the water from escaping the shower? Not necessarily; you can try doubling up with two cotton shower curtains and remembering to shift them to the inside of the tub after your shower to prevent dripping onto the floor. And if you think about it, most hotel shower curtains are cloth and the bathrooms don't seem to get too wet. Hardly a big effort! And as a result of your efforts, you are free from any future worry about toxic fumes emanating from your shower curtain. And an added bonus—cloth curtains last a really long time because when they do start to get icky, you can throw them in the washing machine and, presto, they're like new again. Similar to cotton and maybe even better are hemp shower curtains because they are a bit more mildew resistant than cotton.

Safer Alternatives to PVC 🔋**:** If cotton and hemp don't appeal to you (or maybe you live in a very humid climate), there are safer plastic options than the super-toxic PVC shower curtains described in the report above. Two of the better choices, both of which are widely available, are polyethylene vinyl acetate (PEVA) and ethylene vinyl acetate (EVA). Evidence suggests that both PEVA and EVA are less toxic than traditional PVC shower curtains due to their lack of chlorine. Less easy to find, but also available are HDPE or number 2 shower curtains. All you need to do is look for the letters on the packaging!

Soaps, Shampoos, Deodorant, and Dental Products: Growing up, I don't remember body wash being a thing. Shampoo came in bottles. Sure. But soap came in bar form, not a plastic bottle. Nowadays, everyone and their brother seems to be using soap in a bottle. And I'm only exaggerating slightly. Since body wash was introduced, the sale of bar soaps has fallen 40 percent, and in 2009, body wash outsold bar soaps for the first time ever.

This means there are whole bunch of plastic bottles being used that weren't around just a few short years ago. There's a big difference in the amount of plastic required to package a bar of soap (if any—as a lot seem to come wrapped in paper) versus that bottle of body wash. Of course body wash bottles can be recycled, but oftentimes they wind up in the trash anyway. And even if the bottles are recycled, body wash loses out to bar soap in some other less obvious ways. For one, bar soaps don't contain all the water that body washes do, which means they weigh a lot less. Bars of soap are also smaller, which means less you can fit a lot more of them in a truck. Combining these means less energy used in transporting bar soaps to market as well.

Go Back to the Bar 🍶: If you want to cut back on your plastic bottle usage, reverting back to the good old bar of soap is a great way to go. Oh, and if you're worried about bar soaps being a pathway for sharing icky bacteria or spreading germs, don't be, because a study published in the journal *Epidemiology and Infection* says otherwise. I personally like Dr. Bronner's bar soaps, which are made from natural ingredients, wrapped in paper rather than plastic, reasonably priced, and fairly widely available.

Shampoo and Conditioner 🍶🍶: Shampoo and conditioner is sort of the opposite story. It seems like they have always come in bottles and for the most part still do. But there are plenty of ways you can eliminate shampoo bottles, too. Bars of shampoo are a little trickier to find than bars of soap, so you might have to expand your search beyond your regular grocery store, depending upon where you shop. Nevertheless, the search won't be too difficult because Whole Foods definitely carries some options in this department, so even if you prefer to do your food shopping elsewhere you can still get your plastic-free shampoo there.

Even better than trying Whole Foods, you can try your local natural foods store or co-op. If you don't have any of these options, you could try asking your local store to carry the product for you, or you could buy from a company like Lush: www.lushusa.com. Lush makes both shampoos and conditioners in bar form. If you go into the store, they can cut you a piece of either and sell it to you by weight. You can also order their products online, and they will be delivered wrapped in paper, not plastic.

Deodorant 🔋 🔋 🔋: Prior to writing this book, I'd really not given much thought to the existence of plastic-free deodorant choices. The plastic deodorant case thingy was just so synonymous with deodorant in my mind, I hadn't even thought to try to avoid it. In fact, this may be the final holdout in my slow realization of the truly ubiquitous extent to which plastic is ingrained in our modern lives. But now that I've seen the light, so to speak, it's time to change. Unfortunately, to go plastic free for this one you'll have to search a little bit for options or purchase online.

One thing that is easily available everywhere is good ole baking powder, which some people swear by, but I have yet to try, although I think I will very soon. Supposedly you just go ahead and apply the baking powder to your underarms and you're good to go. I've also heard sage oil works well.

Lush, that company I mentioned in the section on shampoo and conditioner, also sells some solid deodorant, which comes in a block sans the plastic delivery stick in a range of scents. If you are insistent that you want to stay with your original deodorant brand, there still might be options for improvement on the plastic front. Something I'd be willing to bet that few of us have ever considered, but is apparently possible, is recycling deodorant tubes. This possibility never crossed my mind until writing this section of this book. There is something about them that seems unrecyclable, but that isn't true. It's not the most straightforward effort, but it can be done via several potential avenues.

The tricky thing about this is that not all deodorant sticks are created equal—numbers 2, 4, and 5 plastic are all commonly used. That would be HDPE, LDPE, and PP, respectively. And to further complicate things, the little dial on the bottom may be made from a different type of plastic than the rest of the tube.

So what does that mean? It means that you may be required to do a little detective work to determine if you can recycle your tube. Start by taking a peek at the bottom of the tube and see if you can identify the number. Then check with your local recycling center to be sure they will take a deodorant tube of that composition. If that doesn't pan out, there are still other options. The recycling gurus at Earth911 turned me on to the organization TerraCycle. TerraCycle is a great idea. They allow you to box up and mail in hard-to-recycle items, which they upcycle into new products, like really cool tote bags, and sell. They are so very worth checking out. They accept all sorts of Tom's of Maine products, including deodorant sticks. So you could elect to go this route. Unfortunately, it seems this might be the only kind of deodorant stick they currently take, but I'd check back because they seem to be trying hard to do more.

Toothbrushes 🧴 **and Toothpaste** 🧴 🧴 🧴: Options don't abound here, but they do exist. And who knows, maybe by the time you are reading this there will be even more.

Personally, when it comes to toothbrushes I think the Preserve recycled and recyclable (how sweet is that?) toothbrushes are pretty fantastic. We use them exclusively in our household. There really are so many great things about these. First, they are made from recycled yogurt cups, and second, and perhaps best of all, they come in a pouch that you can toss in the mail when done and they will be handily recycled by the company. Oh, and they're BPA free. The package even says you can put two in the package and mail them back, or better yet, put six of them in a plastic water or soda bottle and they will recycle the bottle and the brushes. If you can't find them in your grocery store or at your pharmacy, you can actually subscribe online and they will be sure you get nice clean toothbrushes on a schedule of your choosing. Just don't forget to mail the old ones back.

Other options for toothbrushes exist as well, including ones made from bamboo. You'll just have to shop around.

Toothpaste is a bit trickier. Having spent quite a few years in Maine, we were big fans of Tom's of Maine, and until not that long ago their paste came in aluminum tubes that could be recycled, but not anymore. Unfortunately, now it's plastic. There are still options, just not quite as easily available and recyclable as Tom's. You can also always try to make your own toothpaste, which isn't hard and works quite well. I stumbled upon a recipe online that we quite like, and I must admit, much like the original source claims, it does really get the teeth white. If you want to try it for yourself, here it is:

TOOTHPASTE RECIPE

½ cup baking soda

¼ cup 3 percent hydrogen peroxide

Peppermint oil to taste

Stevia extract to taste

Mix it all up and store it in a jar. We use a mason jar. Oh, and give it a shake before using.

Some sellers, like the good people at Aquarian Bath, also make tooth powders in a tin, and then you can buy refills in a non-plastic bag. You can search their Web page to place orders or just check out their products. If it turns out that homemade toothpaste or tooth powders aren't for you, then at least go for the recycled and recyclable toothbrush. It's very simple, and they work oh so well. Also, if you're bored, maybe tell Tom's to go back to aluminum.

Plastic and Personal Hygiene Quick Tips

1. Beware PVC shower curtains. They release some pretty nasty chemicals. Settle on a safer option.
2. If you've switched from the bar of soap to the bottle of body wash in the last few years, switch back.
3. Lose the tube and make your own toothpaste. It's easy.
4. Use that homemade toothpaste on a recycled and recyclable toothbrush.
5. Ditch the bottle for the block. Try some solid shampoos and conditioners.

Sex Toys: Well, now that you're all showered up and looking great, maybe it's a good time to talk about sex. Turns out sex toys are really, really dirty, and no, I don't just mean in the fun way. Sex toys can fly under the radar because most of them are classified as "novelties," allowing them to forgo regulation because "novelty" implies they won't actually be used. This isn't some tiny industry and a little loophole, but rather an industry valued at fifteen billion dollars by some accounts, suggesting that quite a few people are being exposed to toxins via this avenue. The "novelty" loophole and subsequent lack of regulation allows the manufacturers to use cheap raw ingredients that end up transferring toxic chemicals directly into your body or the body of your partner. Not cool.

Avoid PVC Plastic Sex Toys 🍼🍼🍼: The majority of sex toys are made from polyvinyl chlorides (PVC) or plastic number 3, and many of them are then softened using phthalates. Phthalates have serious potential health effects, including cancer and reproductive system issues like sperm damage. (See chapter 10 for more information on phthalates.) Because cheap plastics are used, the concentrations of toxins like phthalates can be extremely high. One health advocacy group found some sex toys had phthalate concentrations as high as

243,000 ppm, which they deemed as "off the charts," and others found that sex toys off-gassed high levels of toxic phthalates.

While it's not likely the government is going to step in to regulate toxins in sex toys any time soon, it's becoming easier to find safer, less toxic alternatives, especially as celebrities like Oprah and Alicia Silverstone have begun to bring this issue to light and consumers have become more educated. Materials like silicone, metal, and glass are far safer than plastics when it comes to sex toys. Companies like Swan offer toxic-free and rechargeable fun in shapes inspired by nature, and they work great. A quick Google search for eco-friendly sex toys will lead you to more options and bring a whole new meaning to safe sex!

16.

PLASTIC AND YOUR CHILDREN

So now that we've covered sex toys, the next logical step seems to be babies.

Babies are tiny and cute. They are also innocent, vulnerable, and incapable of making their own decisions. As parents, grandparents, aunts, or uncles, we owe it to them to make the most informed decisions possible on their behalf.

This includes decisions about plastics, because *a lot* of baby products are plastic. A real lot. Some of these products are designed to go in babies' mouths and some are not. But I'm not so sure that matters, because most kids put everything in their mouth anyway. A lot of plastic baby products are disposable throwaway crap designed with a short product lifespan in mind. With all that plastic in bottles, toys, and pacifiers, we should ask the question—can these things harm our children? And what are the costs to our environment? If they do contain toxins, can we mitigate the risk to our children while lessening the environmental impacts? Of course we can! It just takes a little know-how.

The remainder of this chapter includes information on common plastic baby products, the problems they present, and some solutions and better options for each. Generally speaking, though, the best way to cut your baby's footprint (plastic and otherwise) is to purchase or use secondhand baby stuff. Not only will you save money, and lots of it, but I believe you will also find buying secondhand likely won't limit your options very much because there is so much baby stuff on craigslist. Some people worry about secondhand baby items exposing their children to germs. Keep in mind that, while it seems somewhat counterintuitive, exposing your children to germs at an early age will likely make them healthier in the long run.

Oh, and just so you know that I'm invested in this section— our daughter will be roughly six months old when this book comes out!

> The ubiquitousness of phthalates in items used daily by children is of concern for children's health because it increases the likelihood of exposure.

Diapers: Over the course of his or her childhood your baby will use an estimated eight thousand diapers! Across the globe as many as 1.375 million diapers are used a day, according to some estimates. Another figure I saw suggested that eighteen billion diapers a year were used in the United States. The EPA says that in 2009 disposable diapers accounted for 3.8 million tons of waste in the municipal waste stream! Nevertheless, 96 percent of babies in America wear disposable diapers. It's apparent that disposable diapers create an *enormous amount of waste!* And that waste can stick around for a really long time, around five hundred years, many generations after we and our babies have departed.

The average baby spends twenty-four hours a day for 2.5 years in diapers. As if all that waste isn't enough to make diapers a worrisome subject, we've no doubt also got to wonder exactly what wrapping my little one in disposable plastics for years might expose her to?

Well . . . there is some evidence linking conventional disposable diapers to potential negative health effects, specifically respiratory problems. A study published in the *Archives of Environmental & Occupational Health* found that disposable diapers off-gassed volatile organic compounds (VOCs), including toluene, ethylbenzene, xylene, and dipentene, which have been linked to cancer and brain damage with long-term or high-level exposure.

The same study also found that mice exposed to chemicals emitted from disposable diapers were more likely to experience negative respiratory effects than those exposed to chemicals from cloth diapers. Bottom line: Some disposable diapers emit chemicals that are toxic to the respiratory tract and may cause or exacerbate asthmatic conditions. Not exactly the type of thing I want to submit my baby to.

Another concern I saw repeated in my research was the potential for diapers to expose babies to dioxins. It's an understandable fear because dioxins are extremely toxic. According to the EPA, dioxins are a group of toxic chemical compounds that cause cancer and are very slow to break down in the environment. Dioxins form as a by-product of the bleaching process in diaper production. They also pose serious health effects at ultra-low concentrations. But what's the deal when it comes to diapers? Are they a harmful source of this awful chemical compound? A 2002 study in *Environmental Health Perspectives* found exposure to dioxins via food is 30,000 to 2,200,000 times greater than exposure through diapers.

Just because disposable diapers themselves don't have high levels of dioxins in them doesn't mean dioxins aren't generated during their production. Because they are. And those dioxins make their way into the environment, where they accumulate in the fatty tissues of other animals that you might one day consume. In the end, the fewer dioxins we release into the environment the better.

Cloth Diapers: So what options exist to reduce waste and protect our children from potentially harmful chemicals? It seems the best option is cloth diapers. For many, even the idea of cloth diapers likely conjures up nightmarish images of safety pins everywhere and poopy cloth tied in knots and falling off of your little one. But, really, that's no longer the case! Today there exist a number of really cool options for reducing waste and keeping your baby happy and clean. No longer is there a need for safety pins, as most modern cloth diapers have Velcro or snaps to hold the diaper on.

Want your cloth diapers to be organic? Sure, no problem. There are quite a few companies making such items. Plus, not only are cloth diapers better for the environment, they are better for your budget. I was unable to find a single comparison suggesting disposable diapers were cheaper, even when factoring in the cost of laundry. And the estimate in cost savings can be huge. I'm talking like two thousand dollars a year! And that's something that I think all parents can agree is a big savings when it comes to baby.

Disposable Hybrid Diapers 🍼🍼: Should you decide that cloth diapers aren't your choice, you can still make a better choice than the conventional disposable. Nowadays, you can also go the disposable hybrid or biodegradable routes. Disposable hybrids are essentially a reusable outer shell with a thin disposable insert. You end up creating less waste, and some of them are biodegradable and can be flushed down the toilet for potential recovery and composting by your municipality or even in your home compost. You can also use a cloth insert instead of a disposable diaper when you're at home. This option seems to afford some flexibility and may be a better environmental option than the conventional disposable route. Especially if you use the disposable insert only on special occasions (e.g., traveling). Other companies are now producing more eco-friendly and baby-friendly options that are free of dyes, chlorine, etc.

Making Cloth Diapers More Efficient

♻ Line dry outside whenever possible.
♻ When replacing appliances, choose more energy-efficient appliances.
♻ Wash with cooler-temperature water.
♻ Wash in larger loads.
♻ Use the same cloth diapers for multiple children.

Diaper Decisions

1. Cloth diapers—No diapers would obviously be ideal, but cloth diapers are the next best thing. Less waste, less money, and fewer toxins than conventional disposables, hybrids, or biodegradables.
2. Hybrids—They aren't ugly, but they aren't good either. Hybrids and biodegradables may alleviate some of the problems with waste and toxins that accompany conventional disposables.

> **3.** Conventional Disposables (Boo!)—I classify these as the worst-case option because of massive amounts of unnecessary waste, higher cost, and even potential negative health effects.

Baby Bottles—BPA Is Banned, but Are They Safe?

The FDA banned BPA from baby bottles and sippy cups in 2012. While this may sound like a progressive and reassuring gesture, it was in reality a pretty hollow one. The truth is, prior to the FDA ban companies had already stopped using BPA in bottles and sippy cups, making the FDA's gesture backward-looking. Whether or not the FDA was late to act, it would seem the end result would be the same—BPA was banned. Problem solved. I wish this was the case. I really do. It would be nice if we could eliminate risks to our children so effectively and clearly.

The ban doesn't pertain to the containers that formula is sold in. So there is certainly potential for the leaching of BPA on that end, unless you live in Connecticut, as they banned BPA in baby formula containers. The United States as a whole has yet to take that step, but it may be coming in the not-so-distant future given the amount of public concern. Until then, it may be best to purchase liquid infant formula sold in glass bottles or even buy the powdered kind, both of which will help limit your baby's exposure to BPA.

However, before we all happily skip down to the baby store, purchase plastic bottles, and feed our kiddos from them, we should think back to a few questions mentioned earlier in this book. Questions like: How BPA free is BPA free? And if there's no BPA, has it been replaced by a similar chemical?

Unfortunately, removing BPA from plastic doesn't mean you remove the effects of BPA. A 2011 study published in *Environmental Health Perspectives* found "almost all commercially available products we sampled—independent of the type of resin, product, or retail source—leached chemicals having detectable EA, including those labeled as BPA free." Oh, darn. Not so simple at all. Some of this EA (estrogenic activity) is likely a result of BPA being replaced by BPS, even in baby bottles. Right now we don't know a whole lot about BPS, but what we do know suggests BPS and its buddies BPB and BPF possess acute toxicity, and EA similar to BPA.

Glass bottles 🍼**:** If you want to be sure your baby isn't exposed to BPA, BPS, or other weird and possibly toxic chemicals via their bottle, the safest route is to buy glass bottles. You might be thinking glass bottles sound like a giant pain in the ass, but they aren't. Why? Because my mom told me so. She said I was raised on glass bottles and that they worked great. Remember, Mom knows best! Seriously though, some really great options exist when it comes to glass bottles. Nowadays you can get them with an external silicone shell that helps prevent

breakage. Plus, when your kid gets older you can swap
out the nipple-shaped top for a regular bottle top and
use it as a BPA-free container to put their lunch bev-
erage in.

Pacifiers: It's hard not to wonder if the plastic in pacifiers is
harmful to your child. After all, they suck on them for hours
at a time. Phthalates were sort of banned from pacifiers and
teething toys in 2008. I say *sort of* because if you dig a little
deeper you'll discover it's not quite so straightforward: Congress "banned three types of phthalates: DEHP, DBP, and
BBP1 in any amount greater than 0.1 percent (computed for
each phthalate individually) in: (1) children's toys, and (2) any
child care article that is designed or intended by the manufacturer to facilitate sleep or the feeding of children age 3 and
younger, or to help children age 3 and younger with sucking
or teething." So I guess this means they aren't truly banned,
but rather limited, which isn't comforting given the extreme
toxicity of phthalates.

BPA isn't banned from pacifiers (in the United States at
least), but it seems that most pacifiers do not contain BPA. Unless it's in the hard outer ring as opposed to the softer nipple
intended to go into your baby's mouth, which is made of latex
or silicone.

Get Some Rubber Pacifiers 🍼: When it comes to
children's health, it's better safe than sorry, so when
our child is ready for a pacifier we'll be purchasing an

all-natural rubber pacifier. Then we can be assured there are no phthalates or BPA anywhere in her pacifier.

Babies don't stay babies forever. At some point your little one is going to transition from infant to toddler to schoolchild and so on. As kiddos undergo these transitions, the products they use and foods they eat will change along with them, and so too will the potential pathways for creating plastic waste and being exposed to toxic chemicals. An interesting stat: The United States has less than 4 percent of the world's children and consumes 40 percent of the world's toys. That's a seriously skewed number, and we're raising some seriously greedy little munchkins. All this consumption is reflected in the enormous sales of the U.S. toy industry, a figure that reached 21.8 billion dollars in 2011. I have no specific stat to reference here, but it's pretty obvious if you walk into any toy store that the majority of these toys are made predominately from plastics. And anyone who has wandered through a big-box store recently or has been around kids' toys knows that these toys aren't built to last from one child to the next, let alone from generation to generation. When these toys break, as they inevitably do, they're more or less bound to end up in the landfill. Recycling toys is extremely difficult because they're often constructed using several types of plastic, making them problematic or impossible to recycle in most facilities.

Think Before You Buy 🍼**:** In researching this section, I stumbled upon a great blog post by a fellow Coloradan. Her post, titled "Where do your kids' toys go to die? Children, consumerism, toys and trash," provides some excellent guidance on cutting the garbage footprint associated with kids' toys. In her post she suggests asking yourself a series of questions before making purchases, which I've modified slightly here:

1. Do I really need this? Does my child really need this?
2. Is this item of good enough quality that it will last for years? Is this item durable enough that someone else can use it when my children are finished with it?

3. Should I save my money for a little while longer and buy a better quality item that will last longer and can be handed down, sold, or donated?
4. Can I save money and resources by purchasing this item secondhand?

The post is also realistic in its expectations and recognizes that not everyone will always act according to these standards, but it reminds us that if you can increase thoughtful purchases by 25, 50, or even 75 percent, think of the huge reduction in cheap plastic toy waste. I think from my perspective the motivation is double. By purchasing less cheap plastic toys, not only do you save on waste, but you'll also save some of your share of the roughly $22 billion we spend on toys each year to spend on your child's college education or help feed, clothe, and educate less-fortunate kids around the world.

The same government law that limits phthalates in baby bottles, pacifiers, and teethers also keeps phthalates out of children's toys. A children's toy as defined by this law as a "consumer product designed or intended by the manufacturer for a child 12 years of age or younger for use by the child when the child plays." A law that protects children under the age of twelve from phthalates sounds like a really great idea, and it is, but not everything is a toy, leaving some pretty big loopholes.

School Supplies: A particularly concerning (and also offensive) pathway for toxins to enter your children's bodies is through their school supplies. I hate the thought of sending kids off to school lugging around a bunch of toxic plastic crap. Law is all about language, and detail and definitions dictate the extent of a law's reach. The language in this law pertains to toys as defined above. School supplies aren't toys because they aren't intended for play, and therefore they're not regulated with respect to phthalates. I'd like to think that despite this loophole the companies manufacturing our children's school supplies would police themselves and use only safe materials . . . but history reveals this is rarely the case.

A report by the Center for Health, Environment & Justice (CHEJ) found school supplies had phthalate levels far exceeding the federal limit for toys. In this report, researchers tested products including The Amazing Spider-Man and Dora the Explorer backpacks only to find they contained phthalates in concentrations up to 69 times the limit banned in toys!

While this is certainly disturbing, it's not illegal. Since it's not illegal, it's up to parents (and teachers) to become educated and take the steps necessary to lessen children's exposure to these endocrine disruptors. Fortunately, there are some pretty simple rules for reducing exposure.

Avoid Vinyl or PVC Supplies 🍶🍶: Do you know how to identify PVC? Look for a V, PVC, or the number 3 in the center of the recycling arrows. If you can't find the number on your school supplies, there are some other tricks to help avoid the worst of the worst.

1. Avoid PVC, aka plastic number 3 or sometimes just plain V. Also avoid PS and PC. These are plastics number 6 and 7, respectively.
2. Shiny plastic backpacks often contain toxins.
3. Avoid notebooks with plastic covers.
4. Watch out for vinyl raincoats, like the traditional yellow ones. They are often loaded with toxins.
5. Use cloth lunch bags instead of plastic lunch boxes.

Lunchtime (for Your Kids or Yourself): One of the first things that comes to mind when I think of school and schoolchildren is packing lunch. But in this age of processed single-serving plastic, how do you make a healthy and less wasteful lunch for yourself or your little ones?

It's easy to picture the more wasteful routes—you know the ones I am talking about: sandwich in a throwaway plastic bag, potato chips in a throwaway plastic bag, cookies in a throwaway plastic bag, all lumped together in a throwaway paper lunch sack. Or maybe you have a plastic lunch box, containing little plastic Baggies and a plastic thermos full of juice. A plastic thermos made of possibly leaching plastics. Or perhaps you have prepackaged, individually wrapped items like juice boxes, Lunchables, Snack Packs, fruit snacks, and cookies. Whether you're wrapping these items yourself or buying them prepackaged (because these things all come served up and wrapped in individual servings nowadays), it's pretty obvious that packing a lunch can create a whole lot of waste.

But whatever are we to do? Do good options exist or am I

condemned to choosing either a wasteful lunch or some expensive time-consuming alternative?

When it comes to cutting plastic waste from packed lunches, a ton of easy and great options exist and they will even save you money! It's a wonderful thing eliminating disposable (or all) plastics from your lunch because it's truly simple. The principle is the same whether you are packing for yourself, your wife, your husband, or your kids, and there are tons of choices no matter whom you're packing lunch for.

Trading in the Plastic Lunch Box and Paper Lunch Bag 🍾: There was a time when lunch boxes were metal. Then they became plastic, but there's no reason you can't go back to metal again. There's also no need for all those disposable paper lunch bags, as you can buy great reusable cloth ones that last a really long time (you can even find organic cotton versions). If you do use a plastic lunch box, be sure to avoid PVC and polystyrene, as they are some of the less friendly plastic options.

No More Plastic Baggies and Skip the Tupperware, Too 🍼**:** Cutting back, or better yet, cutting out single-use plastic sandwich bags is the best way to reduce your lunch's plastic footprint. It's also incredibly easy, basically on par with bringing your own grocery bags to the store. All it takes is the tiniest bit of effort (and money) on the front end. And don't worry about the money; you'll recoup it, no problem. The effort is truly minimal, too. You just need to pick up a reusable alternative, and there are oh-so-many to choose from.

We use reusable cloth bags combined with a set of Pyrex containers and this works great. You can get a set of different-sized ones perfect for things like sandwiches, salads, or leftovers for less than thirty dollars. Consider that a bag of Ziploc sandwich bags goes for around six dollars and you'd likely use three or so a day (sandwich, chips, snack) for about five days a week. That means for each person in your family who makes the switch, you can eliminate around 780 pieces of plastic waste per year, just in sandwich bags! And you'll never have to worry about making lunch and realizing you are out of sandwich bags again. Worse comes to worst, you might need to wash a dirty container so you can pack your lunch. Pyrex isn't the only choice. You could use Tupperware or stainless steel containers (both of which are also available in a bunch of shapes and sizes at a reasonable price), but neither of them is quite as choice as Pyrex, and here's why—

Pyrex is reusable, microwavable, *and* free of toxins. Neither Tupperware nor stainless steel can live up to that claim. Obviously, warming up leftovers stored in stainless steel containers is out the window, and heating up foods in plastics (especially if they are not labeled "microwave safe") is something best to be avoided. By now you are well aware of the dangers associated with heating up foods in plastics.

Stainless steel does have one great advantage, however, especially when it comes to packing lunch for kids. The advantage is that, like plastic, stainless steel containers aren't going to break when your kids drop them, and we all know kids drop stuff all the time. Plus, kids at an age where you need to worry about them shattering glass don't generally microwave their lunch. So win-win.

Another benefit of having an assortment of Pyrex or stainless steel containers around the house is that you can use them as an alternative to plastic bags when you store food in your fridge, and in the process eliminate yet another stream for generating plastic waste and save more money.

Forget the Plastic Thermos and All Those Juice Boxes : Four billion juice boxes get thrown out each year. Four billion! Depending on where you live, you may or may not be able to recycle your children's juice boxes. Juice boxes are difficult to recycle because they are made from layered materials, typically paper (74 percent), plastic, typically polyethylene (22 percent), and aluminum (4 percent). A mix of materials like those found in juice boxes makes for some interesting challenges when it comes to recycling; it also means that a juice box will take as long to decompose in the environment as the slowest decomposing material used in producing it. In this case, that would be plastic. As a conservative estimate, juice boxes are able to persist for three hundred years. And while juice box recycling is difficult, it's important to remember that difficult and impossible are not the same thing. The recycling know-it-alls over at Earth911 explain that because of some efforts by the manufacturers who formed the Carton Council and worked to increase recycling, carton recycling has increased from 18 percent of households in 2005 to 37 percent in 2011. Just check with your local recycling program to be sure! As good as recycling is, not using it in the first place is even better, and the easiest way to avoid juice boxes is to fill your kids up a water bottle or thermos at home—but don't forget that these mediums are also a great way to get some extra phthalates or BPA into your child's body. So be sure to stick to glass or stainless steel thermoses.

Avoid the Single-Serving Plastics and Have Some Healthier Lunch Foods 🍶🍶: This isn't the first time I have made this point, but this is a great place to do so again considering the amount of overly packaged, overly processed crap food marketed to kids these days (adults, too). Next time you're shopping for your kids' lunch, do them (and everyone else) a favor and rather than buying prepackaged highly processed foods, go for something like actual fruit or real cheese. I say everyone else because if we as consumers continually choose healthier, less heavily packaged products, corporations may begin to favor the production and sale of better products. Lastly, while your children might complain a bit at the time, they will thank you later when they are slimmer and more healthy. Because even the Presidential Task Force on Childhood Obesity has acknowledged the role chemicals like BPA may play in fattening up junior. And if the chemicals don't do it, the junk food itself most certainly will.

17

PETS

Anyone who has a dog, myself included (good boy, Hank!), has probably come to realize our furry friends can generate a substantial amount of plastic waste in the form of good old poo bags. The Humane Society says there are about seventy-eight million dogs in the United States. So, let's see, seventy-eight million times two or three poops a day . . . well, you can just imagine the staggering number of plastic bags that wind up in our landfills due to pet waste. Even if not everyone is picking up their dog's poo in plastic bags, you can bet the number of plastic bags used for this purpose each day is in the millions!

Dealing with dog poo: The desire to cut the amount of plastic bags used in cleaning up dog poo leads you to endless options, but none of them are necessarily great cure-alls. Despite that, there are plenty of things that can be done to lessen Fido's plastic footprint, and I am going to run through lots of them, even the ones I find to be not so great.

While we are indeed talking about plastic bags here, the solution isn't so simple, as with plastic grocery bags. I suppose it could be just me, but I don't think reusable dog poop bags would be a particularly huge seller. And like I mentioned before, paper bags are not a super-ecofriendly option either. And realistically, I think picking up dog poo with paper bags would be just a bit more icky. Oh . . . what to do? What to do? Perhaps the best and most obvious solution is to forgo the bags altogether and try and use a pooper-scooper as much as you can. While not exactly a reusable grocery bag or your hands, it serves the same purpose by eliminating the need for single-use bags and the need for material to cover your hands. But what if you don't have a yard? Is a pooper-scooper still relevant? This question raises a good point, which is that depending upon where you live the best options for solving the poo problem may vary considerably. Thus I'm going to lay out some potential solutions based upon a few considerations:

Have a Yard, Scoop the Poop 🍼: Having a yard makes things so, so, so much easier. Especially if you live in a semirural or even suburban place that has a little plot of unmaintained woods. If that's the case, I say train your dog to go to the bathroom in the woods and call it good. If you have a small fenced yard, then the pooper-scooper can be a pretty great method. You can just walk around, pick it up, and (a) compost it, (b) throw it away. Composting is by far preferred. Just remember that you don't want to put compost containing pet waste on vegetables. Instead save it for flowers or other ornamentals. If you are new to composting pet waste, your city doesn't take care of it for you, or you want to learn more, the USDA has a fantastic presentation on how to go about it which I have included a link to in the resources at the end of this book. And that is that. Like I said, have a yard, buy a pooper-scooper.

If you don't have a yard or fenced-in area, things can get a little more complicated. For instance, our apartment is in down-town Boulder, and while it does have a small yard, it's not fenced and it's all xeriscaped (covered in stones and mulch rather than grass), which just isn't ideal for a dog to go outside and poop in. So when Hank needs to go out, we usually take him for a walk to the park, and this requires the ability to pick up his poo. Here are our options, from most effort and most icky to least effort and least icky. These are somewhat inversely related to their footprint.

Reuse Materials You Can't Help But Have 🍼 🍼: What I mean here is, rather than using bags you purchased to pick up your dog's poop, use materials you can't avoid, like circulars that get delivered along with your junk mail and even your actual junk mail. This suggestion came from some readers of my Grist post on this topic. They suggested placing the newspaper or junk mail under your dog while it's pooping and then twisting the ends and disposing of it. I keep wondering if these people have really little dogs that they hold over newspapers while they poo, because this would never work for my dog. Plus, wouldn't you just wind up throwing away newspaper or junk mail that would otherwise have been recycled? So where exactly is the benefit?

Use a Better Bag 🍼: Use a bag that doesn't persist in the environment for a really long time. Ideally, this bag would be one you would bring home and then compost. Such bags do indeed exist. You can find varying degrees of decompostability. If you want to get the really ecofriendly and more truly biodegradable pet waste bags, it seems that going online might be your best bet. You will want to find bags that meet ASTM D6400 specification for biodegradability and compostability. This specification ensures that "plastics and products made from plastics will compost satisfactorily, including biodegrading at a rate comparable to known compostable materials." If it doesn't meet this standard, it may be a bunch of greenwashing. Don't forget to compost them.

Flush Your Dog Waste 🍼🍼🍼: And no, I don't mean train your dog to use the toilet, although that might be an OK solution. What I am talking about is flushable doggie bags made from polyvinyl alcohol (PVA). You

can pick up your pet waste with them while on a walk, then bring them home and flush them down the toilet. This option is not ideal for those who have septic systems, as it can put extra stress on them and ultimately cost you a lot of money.

Dog Treats: Treats are another avenue of waste for your pet, because of all the plastic packaging.

Visit the Treat Bar 🍼: Petco provides plastic bags for you to take your pet treats home in, but they had no problem with me using one of our cloth grocery bags from home.

You Can Also Make Dog Treats 🍼🍼🍼: I've seen all sorts of simple recipes. Should you decide to buy treats prepackaged in a bag, you could also reuse it to pick up some dog poo later!

18.

OUT AND ABOUT

Receipts—The Single Biggest Path to BPA Exposure

Cash register receipts aren't technically plastic, but I'm going to talk about them here anyway. I feel compelled to bring them up because receipts are a great avenue for getting BPA into your body, much like plastics. A study found shocking results when examining cash register receipts. The leader of the study, John Warner, a former professor at the University of Massachusetts and founder of the Warner Babcock Institute for Green Chemistry, makes a critical point. "When people talk about polycarbonate bottles, they talk about nanogram quantities of BPA [leaching out]," Warner observes. "The average cash register receipt that's out there and uses the BPA technology will have 60 to 100 milligrams of free BPA." And just to clarify: 1 milligram = 1,000,000 nanograms.

And unfortunately, receipt paper doesn't need to be consumed to transfer BPA to your body. Rather, the BPA in thermal receipt paper is loosely bound to the paper itself and easily transferred to your skin and directly into your body. So very much so, one study confirmed, that out of four hundred pregnant women in Ohio, the highest BPA levels were found in cashiers, who had almost 20 percent higher concentrations of BPA in their urine than other pregnant women. Another study found that the total mass of BPA on a receipt is 250 to 1,000 times greater than the amount of BPA typically found in a can of food or a can of baby formula.

Avoid the BPA contained in receipts 🍼🍼: This is relatively easy: Just decline a receipt whenever possible.

If You Love the Outdoors

Most of your outdoor gear is plastic, because nylon is plastic: tents, climbing harnesses, backpacks. I could go on, because it's not solely the nylon items; there are these: skis, bike helmets, water filters, sunglasses, waders, running shoes, and canoes. These are all plastic or have plastic elements to them. So all the gear that facilitates those hobbies I love, right down to the camera for photographing the beautiful places I go, contains plastic.

How do I reconcile that? Where can I make changes?

I've thought a lot about the alternatives to the use of plastic in sports. There aren't that many. You can use aluminum or stainless steel bottles for water, but you wouldn't want to go for a two-hour trail run carrying a backpack full of them. I don't particularly want to play catch with a metal frisbee either. But that doesn't mean there aren't any steps we can take:

Look for Recycled Fabrics and Goods: Mountainsmith, a Colorado company, makes some of their backpacks from 100 percent recycled fabrics. I have one and it's every bit as good as any other pack I've owned. I believe that quite a few REI packs are also now made from recycled PET plastic (number 1). Patagonia makes fleeces from recycled plastic bottles. They, too, work great. So let's hope for more of this in the future, because these fabrics make sports and outdoor recreation a lot more fun and a lot safer, so ditching them altogether really isn't an option for some.

So you can seek out goods made from recycled materials or buy them secondhand, but where else might you cut back? I'm honestly not sure. However, you can always give back and help protect the places you love by donating time or money.

New-Car Smell

Of course, if you are going to head to the mountains or to the lake for some fun, you'll need a way to get there. Most likely you will be doing this in a car. Which brings us to the question: What is "new-car smell"? I personally don't care for it. I'd rather have a vase of flowers or some freshly baked bread to sniff. I actually find it a bit nauseating, and it kind of gives me a headache. But, as *CBS News* points out, there are a whole lot of people who feel quite the opposite, so much so that there are sprays and air fresheners for sale that try to re-create that smell. But what exactly is it that those sprays and air fresheners are trying to re-create? Turns out that smell might be a very toxic mix of nasty chemicals off-gassing from all the materials used in building the interior of your car.

When investigating exactly what it is that comprises the new car smell, researchers found 275 different chemicals present, including brominated flame retardants, which are added to plastics to make them less flammable, and chlorine used for polyvinyl chloride (PVC). Both of these have been linked to all sorts of horrific health problems and are best avoided.

Fortunately, things are improving and there's a trend toward using less toxic materials in car interiors. Jeff Gearhart of the Ecology Center, the author of the study, states: "Since we first started testing in 2006, we've seen an improvement on average in the vehicles that are in the market." There's still quite a bit of room for improvement, though, with only 17 percent of cars having PVC-free interiors and 60 percent being free of brominated flame retardants. And improvement is good, considering the average American adult spends 1.5 hours a day in the car.

Research the Cars That Use Less Toxins in Their Materials 🍾🍾: The Ecology Center report lists the ten best and worst car models when it comes to toxins, so before you buy, you may want to check it out. You also might want to keep those windows down in a new car, and especially, allow it to air out after it has been sitting in the sun, where interior air temps near 200 degrees Fahrenheit can potentially lead to even greater concentrations of toxins off-gassing from the interior.

And please do yourself a favor and don't try and re-create that new car smell with sprays.

Conclusion

AKA WHAT HAVE I LEARNED?

mentioned at the very beginning of this book that I thought of plastic as an invasive species. Or at least exhibiting similar characteristics. Now a year later, having spent nearly every second of my free time reading and writing about plastic, I still believe that to be true, but I've come to another conclusion as well:

Plastic is your frenemy

Or an enemy who acts like a friend.

Plastic most certainly meets those criteria. In so many instances, plastic acts like a friend, yet at the same time, despite my best efforts, my body probably contains BPA because of all my exposure to it. Even more disconcerting is the BPA that might be in Mary's body right now. How might it be affecting our unborn child? Frenemy, indeed.

The *Urban Dictionary* definition for frenemy is actually quite long and contains some other valuable insights that might be useful in guiding your future relationship with plastic. For instance: "These relationships are worth doing a cost/benefit analysis on. Also, limiting relationships with frenemies . . . is a must." I couldn't have said it better if I tried. When it comes to plastic, we should all be thankful for the hundreds of positive ways it affects our lives every day, making possible the medical advances that cure us when we are ill, the transportation that lets us see our amazing planet, and the technology that keeps us in touch with friends and family over great distances. But in many instances, and there are oh so many, the costs outweigh the benefits. These are the times we need to distance ourselves from plastic.

My hope is that this book opened your eyes and provided you the knowledge to help protect yourselves and your loved ones from the potentially harmful chemicals lurking in plastics everywhere. At the same time, I hope that you'll find ways to eat a bit healthier, create less waste, and make wiser choices about the plastic in your life.

BIBLIOGRAPHY

Introduction

Bettina Wassener, "Raising Awareness of Plastic Waste," *The New York Times*, August 14, 2011, www.nytimes.com/2011/08/15/business /energy-environment/raising-awareness-of-plastic-waste.html?_ r=0.

Michelle Roberts, "Eating Canned Soup Poses a 'Chemical Risk,'" BBC News, 2011, www.bbc.co.uk/news/health-15834072.

Tiange Wang et al., "Urinary Bisphenol A (BPA) Concentration Associates with Obesity and Insulin Resistance," *Journal of Clinical Endocrinology & Metabolism* 97, no. 2 (March 2012): E223–7, doi:10.1210/jc.2011-1989.

David Biello, "Plastic (Not) Fantastic: Food Containers Leach a Potentially Harmful Chemical," *Scientific American*, February 2008, www.scientificamerican.com/article.cfm?id=plastic-not-fantas tic-with-bisphenol-a.

Kate Sheppard, "Friday Downer: BPA Substitute Is Still Bad For You," *MotherJones.com*, January 2013, www.motherjones.com/blue-marble/ 2013/01/friday-downer-bpa-substitute-still-bad-you.

Dominique Browning, "Hitting the Bottle," *The New York Times*, May 9, 2011.

Bryan Walsh, "The Perils of Plastic," *Time*, April 2010, www.time .com/time/specials/packages/article/0,28804,1976909_1976908 _1976938-4,00.html.

Part One: A Brief History of the Brief History of Plastic

"Lifecycle of a Plastic Product," American Chemistry Council, n.d., http://plastics.americanchemistry.com/Life-Cycle.

Michael Lauzon, "Plastics' Colorful Past," *Plastics News* 19, no. 23 (August 6, 2007).

Erik Lokensgard, *Industrial Plastics: Theory and Application*, 5th ed. (Stamford: Cengage Learning, 2008).

Eve Kahn, "Honoring an Inventor's Passion for Plastic," *The New York Times*, February 5, 2010.

Norman Finkelstein, *Plastics* (Tarrytown, NY: Marshall Cavendish, 2008).

Plastics—The Facts 2012: An Analysis of European Plastics Production, Demand and Waste Data for 2011 (Brussels: PlasticsEurope, 2012).

Susan Freinkel, "A Brief History of Plastic's Conquest of the World," *Scientific American*, May 2011, www.scientificamerican.com/arti cle.cfm?id=a-brief-history-of-plastic-world-conquest.

"Society of Plastics Engineers," description on the Web site of Texas State University, n.d., www.txstate.edu/chemistry/people/clubs/ spe-club.html.

Glenn Beall, "By Design: World War II, Plastics, and NPE," *Plastics Today.com*, 2009, www.plasticstoday.com/imm/articles/design -world-war-ii-plastics-and-npe.

Lien Hoang, "Agent G.M.O.," *The New York Times*, March 26, 2013, http://latitude.blogs.nytimes.com/2013/03/26/should-monsanto -be-allowed-to-bring-genetically-engineered-crops-to-vietnam/? _r=0.

"Agent Orange: Background on Monsanto's Involvement," Monsanto,

n.d., www.monsanto.com/newsviews/Pages/agent-orange-back
ground-monsanto-involvement.aspx.

Heather Rogers, "A Brief History of Plastic," *The Brooklyn Rail*, May
2005, www.brooklynrail.org/2005/05/express/a-brief-history-of
-plastic#.

Jeffrey Meikle, *American Plastic: A Cultural History* (New Brunswick,
NJ: Rutgers University Press, 1995).

**2: A Story About Nylon: From Coal, Water, and Air to War and
Riots!**

David Hounshell, *Science and Corporate Strategy: DuPont R&D, 1902–
1980* (Cambridge: Cambridge University Press, 1988).

Emily Spivack, "Stocking Series, Part 1: Wartime Rationing and Nylon
Riots," *Smithsonian.com*, 2012, http://blogs.smithsonianmag.com
/threaded/2012/09/stocking-series-part-1-wartime-rationing-and
-nylon-riots/.

Giles Slade, *Made to Break: Technology and Obsolescence in America*
(Cambridge, MA: Harvard University Press, 2009).

Emily Spivack, "Stockings Series, Part 2: Paint-on Hosiery During the
War Years," *Smithsonian.com*, 2012, http://blogs.smithsonianmag
.com/threaded/2012/09/stockings-series-part-2-paint-on-hosiery
-during-the-war-years/.

Gertrude Bailey, "First Year of Peace Will Bring Back Nylons," *The
Pittsburgh Press*, July 18, 1944.

Susannah Handley, *Nylon: The Story of a Fashion Revolution* (Baltimore:
JHU Press, 1999).

3: Tupperware—A Classic Plastic Product We're All Familiar With

"Tupper's Invention Notebooks," PBS.org, 2004, www.pbs.org/wgbh
/americanexperience/features/photo-gallery/tupperware-inven
tions/.

Laurie Kahn-Leavitt, "Tupperware!" (American Experience, 2004),
www.pbs.org/wgbh/americanexperience/films/tupperware/.

"Timeline: Women, Work, and Plastics History," PBS.org, 2004, www.pbs.org/wgbh/americanexperience/features/timeline/tupperware/.

Shellie Karabell, "Tupperware: A Party Somewhere Every Two Seconds," INSEAD.edu, 2009, http://knowledge.insead.edu/leadership-management/strategy/tupperware-a-party-somewhere-every-two-seconds-1390.

Diane Cardwell, "Solar Industry Borrows a Page, and a Party, from Tupperware," *The New York Times*, December 1, 2012, www.nytimes.com/2012/12/01/business/energy-environment/solar-industry-borrows-a-page-and-a-party-from-tupperware.html.

Alison Clarke, *Tupperware: The Promise of Plastic in 1950's America* (Washington, DC: Smithsonian Books, 2001).

Plastics—The Facts 2012: An Analysis of European Plastics Production, Demand and Waste Data for 2011 (Brussels: PlasticsEurope, 2012).

"Per Capita Plastics Consumption (2001 vs. 2010)." *Plastics News.* www.plasticsnews.com/article/20030929/FYI/309299998/per-capita-plastics-consumption-2001-vs-2010

National Institutes of Health and U.S. Department of Health and Human Services, *U.S. Health in International Perspective: Shorter Lives, Poorer Health* (The National Academies Press: 2013), http://sites.nationalacademies.org/DBASSE/CPOP/US_Health_in_International_Perspective/.

Part Two: The Science Behind the Plastic

4: How Is Plastic Produced?

David Biello, "Plastic from Plants: Is It an Environmental Boon or Bane?," *ScientificAmerican.com*, October 26, 2010, http://www.scientificamerican.com/article.cfm?id=is-plastic-from-plants-good-for-the-environment-or-bad.

Erik Lokensgard, *Industrial Plastics: Theory and Application*, 5th ed. (Stamford: Cengage Learning, 2008).

5: How Much Fossil Fuel Is Consumed in the Production of Plastics?

Umbra Fisk, "Umbra on Oil and Plastic," Grist.org, 2007, http://grist .org/article/plastics/.

"Short-Term Energy Outlook," U.S. Energy Information Administration, 2013, www.eia.gov/forecasts/steo/report/global_oil.cfm.

6: The Many Types of Plastic: What Do the Numbers Inside the Little Recycling Symbols Actually Mean?

"Plastics," EPA.org, 2013, www.epa.gov/osw/conserve/materials/plas tics.htm.

"Types of Plastic," *PlasticsEurope.org*, www.plasticseurope.org/what -is-plastic/types-of-plastics-11148.aspx.

7: Recycling Plastic

"Save Energy by Recycling: Full iWarm Tool," U.S. Environmental Protection Agency, n.d., www.epa.gov/osw/conserve/tools/iwarm/.

"Frequently Asked Questions," U.S. Energy Information Administration, n.d., www.eia.gov/tools/faqs/faq.cfm?id=97&t=3.

Leah Blunt, "Recycling Mystery: Milk and Juice Cartons," Earth911 .com, 2012, http://earth911.com/news/2012/01/02/recycling-mys tery-milk-and-juice-cartons/.

"The ImpEE Project," Department of Engineering, Cambridge University, 2005, www-g.eng.cam.ac.uk/impee/resources/.

Rachel Cernansky, "When Recycling Is Bad for the Environment," *Discover*, July–August 2009, http://discovermagazine.com/2009 /jul-aug/06-when-recycling-is-bad-for-the-environment# .UfgUQGQfZU5.

"Dirty Dozen Recycling Contaminants," Eco-Cycle.org, http://eco cycle.org/dirtydozen.

Why is plastic recycling such a mess? Can it get better?

Renee Cho, "What Happens to All that Plastic," State of the Planet Blog, Earth Institute, January 2012, http://blogs.ei.columbia.edu /2012/01/31/what-happens-to-all-that-plastic/.

Bettina Wassener, "Raising Awareness of Plastic Waste," *The New York Times*, August 14, 2011 www.nytimes.com/2011/08/15/business/energy-environment/raising-awareness-of-plastic-waste.html?_r=0."

Anthony L. Andrady and Mike A. Neal, "Applications and Societal Benefits of Plastics," *Philosophical Transactions of the Royal Society of London. Series B, Biological Sciences* 364, no. 1526 (July 27, 2009): 1977–84, doi:10.1098/rstb.2008.0304.

Daniel Robison, "Startup Converts Plastic to Oil, and Finds a Niche," NPR.org, 2012, www.npr.org/2012/03/19/147506525/startup-converts-plastic-to-oil-and-finds-a-niche.

Mike Biddle, "Mike Biddle: We Can Recycle Plastic," YouTube, TED Talks, 2011, www.youtube.com/watch?v=RD07GkmM2fc.

Justin McCurry, "Japan Streets Ahead in Global Plastic Recycling Race," *The Guardian*, December 29, 2011, http://www.theguardian.com/environment/2011/dec/29/japan-leads-field-plastic-recycling.

Sophia Jones, "Sweden Wants Your Trash," NPR.org, 2012, www.npr.org/blogs/thetwo-way/2012/10/28/163823839/sweden-wants-your-trash.

8: Bioplastics: Like the Kid Sister of Biofuels

Mara Lemos Stein and Naureen Malik, "Just One Word: Bioplastics," *The Wall Street Journal*, October 18, 2010, http://online.wsj.com/article/SB10001424052748703989304575504141785646492.html.

"US Bioplastics Demand Growing 20 Percent Annually," *Plastics News.com*, 2012, www.plasticsnews.com/article/20120709/NEWS/307099979/us-bioplastics-demand-growing-20-percent-annually.

Bioplastics Industry Overview Guide, The Society of the Plastics Industry, Bioplastics Council, April 2012, www.plasticsindustry.org/files/about/BPC/Industry Overview Guide Executive Summary -0912-Final.pdf.

John Rather, "Tapping Power from Trash," *The New York Times*, September 13, 2008.

Patrick Sullivan and EEC Research Associate, *The Importance of Landfill Gas Capture and Utilization in the U.S.*, Earth Engineering Center, Columbia University, April 2010, www.seas.columbia.edu/ earth/wtert/sofos/Importance_of_LFG_Capture_and_Utilization_in_the_US.pdf.

IPCC, 2007: Climate Change 2007: The Physical Science Basis. Contribution of Working Group I to the Fourth Assessment Report of the Intergovernmental Panel on Climate Change [Solomon, S., D. Qin, M. Manning, Z. Chen, M. Marquis, K. B. Averyt, M. Tignor and H. L. Miller (eds.)]. (Cambridge, United Kingdom and New York, NY, USA: Cambridge University Press, 2007).

Erica Gies, "Bioplastics: The Challenge of Viability," *The New York Times*, July 06, 2008, www.nytimes.com/2008/07/06/business/worldbusiness/06iht-rbogplast.sr.14263750.html?pagewanted=all&_r=0.

"Sustainability Metrics: Life Cycle Assessment and Green Design in Polymers." Michaelangelo D. Tabone, James J. Cregg, Eric J. Beckman, and Amy E. Landis. *Environmental Science & Technology* 2010 44 (21): 8264–8269.

Part Three: The Good, the Bad, and the Ugly Plastics

9: The Good: Benefits of Plastic

"10 Things We Wish Had Never Been Invented," Brainz.org, http:// brainz.org/10-things-wish-never-invented.

Anthony L. Andrady and Mike A. Neal, "Applications and Societal Benefits of Plastics," *Philosophical Transactions of the Royal Society of London. Series B, Biological Sciences* 364, no. 1526 (July 27, 2009): 1977–84, doi:10.1098/rstb.2008.0304.

David Biello, "Plastic (Not) Fantastic: Food Containers Leach a Potentially Harmful Chemical," *Scientific American*, February 2008, www.scientificamerican.com/article.cfm?id=plastic-not-fantastic -with-bisphenol-a."

Elisabeth Rosenthal, "Your Biggest Carbon Sin May Be Air Travel," *The New York Times*, January 26, 2013, www.nytimes.com/2013/01/27/sunday-review/the-biggest-carbon-sin-air-travel.html?_r=0.

The Plastics Industry Trade Association, "Plastics: Making Modern Life Possible," www.plasticsindustry.org/AboutPlastics/contentwip.cfm?ItemNumber=6785

Ensinger, *High Performance Plastics for Renewables*, www.ensinger-online.com/uploads/media/Renewable-Energies-Ensinger-GB_02.pdf.

10: The Bad: Toxins and Plastic

Bryan Walsh, "The Perils of Plastic," *Time*, April 2010, www.time.com/time/specials/packages/article/0,28804,1976909_1976908_1976938-4,00.html.

Cheryl Erler and Julie Novak, "Bisphenol a Exposure: Human Risk and Health Policy," *Journal of Pediatric Nursing* 25, no. 5 (October 2010): 400–7, doi:10.1016/j.pedn.2009.05.006.

"Endocrine Disruptors," National Institute of Environmental Health Sciences, n.d., www.niehs.nih.gov/health/topics/agents/endocrine/index.cfm.

Pete Meyers, "Chapel Hill Bisphenol A Expert Panel Consensus Statement: Integration of Mechanisms, Effects in Animals and Potential Impact to Human Health at Current Exposure Levels," *EnvironmentalHealthNews.org*, 2007.

Jon Hamilton, "Link Between BPA and Childhood Obesity Is Unclear," NPR.org, 2012, www.npr.org/blogs/health/2012/09/18/161340024/link-between-bpa-and-childhood-obesity-is-unclear.

Daniel R. Doerge et al., "Distribution of Bisphenol A into Tissues of Adult, Neonatal, and Fetal Sprague-Dawley Rats," *Toxicology and Applied Pharmacology* 255, no. 3 (September 15, 2011): 261–70, doi:10.1016/j.taap.2011.07.009.

Frederick S. vom Saal et al., "Chapel Hill Bisphenol A Expert Panel

Consensus Statement: Integration of Mechanisms, Effects in Animals and Potential to Impact Human Health at Current Levels of Exposure," *Reproductive Toxicology* 24, no. 2: 131–8, accessed July 30, 2013, doi:10.1016/j.reprotox.2007.07.005.

De-Kun Li et al., "Urine Bisphenol—a Level in Relation to Obesity and Overweight in School-age Children," *PLOS ONE* 8, no. 6 (January 2013): e65399, doi:10.1371/journal.pone.0065399.

The University of Alabama at Birmingham, "Researchers Examine BPA and Breast Cancer Link," ScienceDaily.com, 2011, www.sciencedaily.com/releases/2011/10/111018214107.htm.

Amy Silverstein, "Is Susan G. Komen Denying the BPA-Breast Cancer Link?," *MotherJones.com*, 2011, www.motherjones.com/environment/2011/09/breast-cancer-komen-bpa.

Julian Josephson, "Chemical Exposures: Prostate Cancer and Early BPA Exposure," *Environmental Health Perspectives* 114, no. 9 (2006): A520, www.ncbi.nlm.nih.gov/pmc/articles/PMC1570083/.

Amanda Harper, "UC to Study Link Between BPA and Prostate Cancer," HealthNews, UC Academic Health Center, 2008, http://healthnews.uc.edu/publications/findings/?/6737/6773/.

Jen Quraishi, "Large Study Confirms BPA-Thyroid Function Link," *MotherJones.com*, 2011, www.motherjones.com/blue-marble/2011/07/large-study-confirms-bpa-affects-thyroid-function.

Kerry Grens, "BPA Chemical May Be Tied to Heart Disease," Reuters.com, 2012, www.reuters.com/article/2012/03/06/us-bpa-chemical-tied-heart-disease-idUSTRE8251KL20120306.

Karen Peart, "More Evidence That BPA Found in Clear Plastics Impairs Brain Function," Yale.edu, September 2008, http://news.yale.edu/2008/09/03/more-evidence-bpa-found-clear-plastics-impairs-brain-function.

Nicholas Kristof, "Big Chem, Big Harm?," *The New York Times*, August 26, 2012, www.nytimes.com/2012/08/26/opinion/sunday/kristof-big-chem-big-harm.html?_r=0&.

"France to Ban BPA in Food Packaging," *PlasticsNews.com*, October 2012, www.plasticsnews.com/article/20121005/NEWS/310059985.

Felix Grün and Bruce Blumberg, "Environmental Obesogens: Organotins and Endocrine Disruption via Nuclear Receptor Signaling," *Endocrinology* 2006, 147 (6): s50–s55.

Raquel Chamorro-García, Margaret Sahu, Rachelle J. Abbey, Jhyme Laude, Nhieu Pham, and Bruce Blumberg, "Transgenerational Inheritance of Increased Fat Depot Size, Stem Cell Reprogramming, and Hepatic Steatosis Elicited by Prenatal Exposure to the Obesogen Tributyltin in Mice." *The Free Library*, March 1, 2013; August 24, 2013, www.thefreelibrary.com/Transgenerational inheritance of increased fat depot size, stem cell . . . -a0332248651>.

Australian Competition and Consumer Commission, "Phthalates in Consumer Products," ProductSafety.gov.au, www.productsafety .gov.au/content/index.phtml/itemId/972486.

Brigham and Women's Hospital, "Chemicals in Personal Care Products—Phthalates—May Increase Risk of Diabetes in Women," *ScienceDaily.com*, July 13, 2012, www.sciencedaily.com/releases /2012/07/120713083103.htm.

Jonathan Chevrier and Kathleen McCarty, "Phthalates Exposure May Double Breast Cancer Risk . . . or Not," EnvironmentalHealth News.org, March 19, 2010.

Tamarra James-Todd et al., "Urinary Phthalate Metabolite Concentrations and Diabetes Among Women in the National Health and Nutrition Examination Survey (NHANES) 2001–2008," *Environmental Health Perspectives* 120, no. 9 (2012): 1307–1313, www.med scape.com/viewarticle/772052_1.

60 Minutes, "Phthalates: Are They Safe?," *CBSNews.com*, 2010.

Jane Kay and Environmental Health News, "Johnson & Johnson Removes Some Chemicals from Baby Shampoo, Other Products," *ScientificAmerican.com*, May 5, 2013.

The Associated Press, "California Bans Plastic Toy Chemical," *CBS News.com*, February 11, 2009, www.cbsnews.com/2100-204_162 -3366238.html.

"Drug Coatings Can Contain Problematic Chemicals," *Discovery News*, December 22, 2011, http://news.discovery.com/human/pill-coat ings-chemicals-hormones-111222.htm.

Sanjay Gupta and Elizabeth Cohen, "Diabetes and Cosmetics: A Connection?," *The Chart* (blog, cnn.com), July 13, 2012, http://thechart.blogs.cnn.com/2012/07/13/diabetes-and-cosmetics-a-connection/.

Dan Shapley, "How to Avoid Phthalates in 3 Steps," thedailygreen.com, February 4, 2008.

Ruthann A. Rudel et al. "Food Packaging and Bisphenol A and Bis(2-ethyhexyl) Phthalate Exposure: Findings from a Dietary Intervention," *Environmental Health Perspectives* 119, no. 7 (July 2011): 914–20, doi:10.1289/ehp.1003170.

R. Hauser and A. M. Calafat, "Phthalates and Human Health," *Occupational and Environmental Medicine* 62, no. 11 (December 2005): 806–18, doi:10.1136/oem.2004.017590.

Styrene and benzene: A whole other reason to hate polystyrene

"Styrofoam: The Eco-Enemy," SierraClubGreenHome.com, www.sierraclubgreenhome.com/featured/pop-goes-the-polystyrene/.

"Styrene (monomer) MSDS" (ScienceLab.com, 2013), www.sciencelab.com/msds.php?msdsId=9925112.

Centers for Disease Control and Prevention, "Facts About Benzene," Bt.cdc.gov, February 14, 2013, www.bt.cdc.gov/agent/benzene/basics/facts.asp.

11: The Ugly: The Environmental Costs of Our Plastic Addiction

V. E. Yarsley and E. G. Couzens, *Plastics* (Middlesex: Penguin Books Limited, 1945).

BBC News, "South Africa Bans Plastic Bags," News.bbc.co.uk, May 9, 2003.

Brian Clark Howard, "Plastic Bag Taxes Don't Hurt Low-Income People," Newswatch.nationalgeographic.com, November 25, 2012, http://newswatch.nationalgeographic.com/2012/11/25/plastic-bag-taxes-dont-hurt-low-income-people/.

Murray R. Gregory, "Environmental Implications of Plastic Debris in Marine Settings—Entanglement, Ingestion, Smothering, Hangers-on, Hitch-hiking and Alien Invasions," *Philosophical Transactions of the Royal Society of London. Series B, Biological Sciences* 364, no. 1526 (July 27, 2009): 2013–25, doi:10.1098/rstb.2008.0265.

The Worldwatch Institute, *The Worldwatch Institute Annual Report 2004*, 2004, www.worldwatch.org/system/files/Annual_Report-2004.pdf.

Jort Hammer, Michiel H. S. Kraak, and John R. Parsons, "Plastics in the Marine Environment: The Dark Side of a Modern Gift," *Reviews of Environmental Contamination and Toxicology* 220 (January 2012): 1–44, doi:10.1007/978-1-4614-3414-6_1.

National Oceanic and Atmospheric Administration, "Frequently Asked Questions: What We Actually Know About Common Marine Debris Factoids," NOAA.gov, August 28, 2012, http://marinedebris.noaa.gov/info/faqs.html#4.

Michelle Allsopp et al., *Plastic Debris in the World's Oceans*, Greenpeace International, 2005, www.unep.org/regionalseas/marinelitter/publications/docs/plastic_ocean_report.pdf.

Renee Cho, "Our Oceans: A Plastic Soup," *State of the Planet Blog*, January 26, 2011.

Paula Alvarado, "Traversing the South Pacific Gyre: Plastic Marine Pollution Is Officially a Global Issue," TreeHugger.com, April 8, 2011, www.treehugger.com/about-treehugger/traversing-the-south-pacific-gyre-plastic-marine-pollution-is-officially-a-global-issue.html.

C. J. Moore et al., "A Comparison of Plastic and Plankton in the North Pacific Central Gyre," *Marine Pollution Bulletin* 42, no. 12 (December 2001): 1297–1300, doi:10.1016/S0025-326X(01)00114-X.

Annalee Newitz, "Lies You've Been Told About the Pacific Garbage Patch," Io9.com, May 21, 2012, http://io9.com/5911969/lies-youve-been-told-about-the-pacific-garbage-patch.

UNEP, *Marine Litter: A Global Challenge* (Nairobi: UNEP, 2009).

Part Four: Time to Purge Some Ugly Plastic

Ruthann A. Rudel et al., "Food Packaging and Bisphenol A and Bis(2-ethyhexyl) Phthalate Exposure: Findings from a Dietary Intervention."

Jennifer T. Wolstenholme et al., "Gestational Exposure to Bisphenol A Produces Transgenerational Changes in Behaviors and Gene Expression," *Endocrinology* 153, no. 8 (August 2012): 3828–38, doi:10.1210/en.2012-1195.

Nicholas Kristof, "Warnings from a Flabby Mouse," *The New York Times*, January 20, 2013, www.nytimes.com/2013/01/20/opinion/sunday/kristof-warnings-from-a-flabby-mouse.html?smid=fb-share&_r=0&.

"Farmers Markets Search," USDA.gov, 2013, http://search.ams.usda.gov/farmersmarkets/.

Chris Conway, "Taking Aim at All Those Plastic Bags," *The New York Times*, April 01, 2007, www.nytimes.com/2007/04/01/weekinreview/01basics.html.

Australian Bureau of Statistics, "How Much Energy Is Used to Make a Plastic Bag?," 2004, www.abs.gov.au/ausstats/abs@.nsf/94713ad445ff1425ca25682000192af2/2498b7e0c5178282ca256dea000539bc!.

City of Olympia, "Ditch the Plastic," Olympiawa.gov, August 21, 2012, http://olympiawa.gov/city-utilities/garbage-and-recycling/zero-waste/ditch-the-plastic.

The U.S. Environmental Protection Office of Resource Conservation and Recovery, *Municipal Solid Waste Generation, Recycling, and Disposal in the United States Tables and Figures for 2010*, 2011, www.epa.gov/osw/nonhaz/municipal/pubs/msw_2010_data_tables.pdf.

Robb Krehbiel, "New Report: Recycling Cannot Solve Plastic Bag Problem," EnvironmentWashington.org, February 14, 2012, www.environmentwashington.org/news/wae/new-report-recycling-cannot-solve-plastic-bag-problem.

Kitt Doucette, "The Plastic Bag Wars," *RollingStone.com*, July 25, 2011, www.rollingstone.com/politics/news/the-plastic-bag-wars-20110725.

Brian Clark Howard, "Plastic Bag Taxes Don't Hurt Low-Income People," newswatch.nationalgeographic.com, November 25, 2012.

Jon Entine, "Campbell's Big Fat Green BPA Lie—and the Sustainability Activists Who Enabled It," *Forbes.com*, September 18, 2012.

Brian Palmer, "Disoriented in the Dairy Aisle: Should I Buy Milk in Glass, Plastic, or Cardboard Containers?," *Slate*, 2011, www.slate .com/articles/health_and_science/the_green_lantern/2011/03/ disoriented_in_the_dairy_aisle.html.

Chris Morran, "We Are Apparently in the midst of a Canned Beer Renaissance," Consumerist.com, June 20, 2012, http://consumer ist.com/2012/06/20/we-are-apparently-in-the-midst-of-a-canned -beer-renaissance/.

Health Canada: Bureau of Chemical Safety, Food Directorate, and Health Products and Food Branch, *Survey of Bisphenol A in Canned Drink Products*, 2009, www.hc-sc.gc.ca/fn-an/securit/packag-emball/bpa/bpa_survey-enquete-can-eng.php.

Lloyd Alter, "More Americans Drinking BPA in Canned Beer, Thanks to Economy and Pabst Drinking Hipsters," TreeHugger.com, June 20, 2012, www.treehugger.com/health/more-americans-are-drink ing-bpa-canned-beer-thanks-economy-and-pabst-drinking-hip sters.html.

Pablo Paster, "Ask Pablo: Do New Six-Pack Rings Offer a More Sustainable Solution?," TreeHugger.com, 2010, www.treehugger.com /green-food/ask-pablo-do-new-six-pack-rings-offer-a-more-sus-tainable-solution.html.

Lori Brown, "Wow, You Can Recycle That?," Earth911.com, February 8, 2012, http://earth911.com/news/2010/02/08/wow-you-can-re cycle-that/.

Mark Baumgartner, "Study: Bottled Water No Safer Than Tap Water," *Abcnews.go.com*, May 03, 2001.

Peter Gleick, *Bottled and Sold: The Story Behind Our Obsession with Bottled Water* (Washington, DC: Island Press, 2010).

Erik Olson, *Bottled Water Pure Drink or Pure Hype?* (NRDC, 1999), www.nrdc.org/water/drinking/bw/bwinx.asp.

Environmental Protection Agency, "Water Trivia Facts," EPA.org.

Mat McDermott, "US Bottled Water Sales Hit New Record High in 2011," TreeHugger.com, May 29, 2012, http://www.treehugger.com /corporate-responsibility/united-states-bottled-water-sales-new-re cord-high-2011.html.

The Pacific Institute, "Bottled Water and Energy Fact Sheet," Pacinst
.org, www.pacinst.org/publication/bottled-water-and-energy-a-fact
-sheet/.

CBC News, "Health Canada Finds Bisphenol A in Soft Drinks," Cbc.
ca, March 06, 2009.

Biello, "Plastic from Plants: Is It an Environmental Boon or Bane?"

Anya Kamenetz, "The Starbucks Cup Dilemma," Fastcompany.com,
October 20, 2010, www.fastcompany.com/1693703/starbucks-cup
-dilemma.

U.S. Environmental Protection Agency, "10 Fast Facts on Recycling,"
EPA.org, October 31, 2012, www.epa.gov/reg3wcmd/solidwaster
ecyclingfacts.htm.

Brock Parker, "Brookline Mulls Ban on Styrofoam, Plastic Bags," *The
Boston Globe*, September 30, 2012, www.bostonglobe.com/metro
/regionals/west/2012/09/29/brookline-mulls-ban-styrofoam-plas
tic-bags/4T5lq7RPYoaXFoW16M0uOI/story.html.

Chun Z. Yang et al., "Most Plastic Products Release Estrogenic Chemicals: A Potential Health Problem That Can Be Solved.," *Environmental Health Perspectives* 119, no. 7 (July 2011): 989–96, doi:10.1289
/ehp.1003220.

J. Chang, R. Fortmann, and N. Roache, "Air Toxics Emissions from a
Vinyl Shower Curtain," *Indoor Air 2002—9th International Conference on Indoor Air Quality and Climate* (Rotterdam: 2002), 542–547,
http://cfpub.epa.gov/si/si_public_record_Report.cfm?dirEnt
ryId=63907&CFID=127785508&CFTOKEN=83092546&jsessi
onid=4e307ba332a89c6c34ee333b5315294943a/.

Center for Health, Environment & Justice, "Volatile Vinyl—the New
Shower Curtain's Chemical Smell?," Chej.org, http://chej.org/
campaigns/pvc/resources/shower-curtain-report/.

Deirdre Dolan, "How to Choose a Safe Shower Curtain," thedailygreen
.com, July 31, 2008, www.thedailygreen.com/living-green/blogs
/organic-parenting/safe-shower-curtains-55073102.

Scott Graf, "Lather Up: More Men Switching to Shower Gels," NPR
.org, July 7, 2010, www.npr.org/templates/story/story.php?
storyId=128341803&ft=1&f=1003.

Carol Thompson, "Use Soap, Not Body Wash, for a Greener Clean," thedailygreen.com, July 30, 2010, www.thedailygreen.com/green -homes/eco-friendly/body-wash-vs-bar-soap-460710.

J. Heinze and F. Yackovich, "Washing with Contaminated Bar Soap Is Unlikely to Transfer Bacteria," *Epidemiology and Infection* 101, no. 1 (1988): 135–142, www.ncbi.nlm.nih.gov/pmc/articles /PMC2249330/.

Virginia Sole-Smith and Planet Green, "The Plastic Project Part 5: Three Natural Deodorant Alternatives That Work," The Learning Channel, Tlc.howstuffworks.com, http://tlc.howstuffworks.com /style/plastic-project-part-5-3-natural-deodorant-alternatives-that -work.htm.

Mary Mazzoni, "Recycling Mystery: Deodorant Tubes," Earth911. com, September 4, 2012, http://earth911.com/news/2012/09/04 /recycling-deodorant-tubes/.

"Solid Shampoo," *type A minimalist* (blog), typeaminimalist.com, August 1, 2011, http://typeaminimalist.com/tag/solid-shampoo/.

Brian Alexander, "When Sex Toys Turn Green—for Health, That Is," *NBCNews.com*, March 21, 2007, www.nbcnews.com/id /19333870/#.UTVKbDB9ZKY.

Hana Alberts, "Pleasure's Stiff Price," *Forbes.com*, December 10, 2009, www.forbes.com/2009/12/10/million-dollar-vibrator-expensive -sex-toy-lifestyle-style-adult-entertainment.html.

Emily Gertz, "Ever Thought About the Toxins in Your Sex Toys?," Grist .org, n.d., http://grist.org/article/gertz1/.

Helen Thompson, "Early Exposure to Germs Has Lasting Benefits," *Nature*, March 22, 2012, doi:10.1038/nature.2012.10294.

U.S. Environmental Protection Agency, *Phthalates: Toxicity and Exposure Assessment for Children's Health*, TEACH Chemical Summary, U.S. EPA, 2007, www.epa.gov/teach/chem_summ/phthalates_ summary.pdf.

U.S. Environmental Protection Agency, "10 Fast Facts on Recycling," EPA.org, October 31, 2012.

"How Many Diapers Are Required Every Day to Satisfy the World

Consumption?," DisposableDiaper.net, http://disposablediaper.net/faq/how-many-diapers-are-required-every-day-to-satisfy-the-world-consumption/.

U.S. Environmental Protection Agency, *Municipal Solid Waste in the United States: 2009 Facts and Figures*, 2010, www.epa.gov/wastes/nonhaz/municipal/pubs/msw2009rpt.pdf.

Kiera Butler and Dave Gilson, "What's Your Baby's Carbon Footprint?," *MotherJones.com*, April 28, 2008, www.motherjones.com/environment/2008/04/whats-your-babys-carbon-footprint.

The University of Michigan Health System, "Toilet Training," Med.umich.edu, March 2010, www.med.umich.edu/yourchild/topics/toilet.htm.

R. C. Anderson and J. H. Anderson, "Acute Respiratory Effects of Diaper Emissions," *Archives of Environmental & Occupational Health* 54, no. 5: 353–8, accessed July 31, 2013, doi:10.1080-/00039899909602500.

Anna Fahey, "The Great Diaper Debate," Grist.org, December 19, 2009, http://grist.org/article/the-great-diaper-debate/.

World Heath Organization, *Dioxins and Their Effects on Human Health*, 2010, www.who.int/mediacentre/factsheets/fs225/en/.

The Editors of *E Magazine*, "EarthTalk: Cloth or Disposable? The Diaper Debate Is Back," *Csmonitor.com*, January 23, 2009, www.csmonitor.com/Environment/Living-Green/2009/0123/earthtalk-cloth-or-disposable-the-diaper-debate-is-back.

Marla Cone, "Long-Awaited Dioxins Report Released; EPA Says Low Doses Risky but Most People Safe," *EnvironmentalHealthNews.org*, February 17, 2012, www.environmentalhealthnews.org/ehs/news/2012/dioxins-report-revealed.

Michael DeVito and Arnold Schecter, "Exposure Assessment to Dioxins from the Use of Tampons and Diapers.," *Environmental Health Perspectives* 110, no. 1 (2002): 23–28, www.ncbi.nlm.nih.gov/pmc/articles/PMC1240689/.

Umbra Fisk, "Umbra on Chlorine," Grist.org, http://grist.org/article/chlorine/.

Laurie Benner, "Just One Thing: How to Green Your Baby's Diapers," *Abcnews.go.com*, March 26, 2010, http://abcnews.go.com/GMA /JustOneThing/thing-green-babys-diapers/story?id=10202689.

The Associated Press, "FDA: BPA Banned in Baby Bottles," *NBC News. com*, July 17, 2012, www.nbcnews.com/id/48209901/ns/health -childrens_health/#.UfmfPGQfZU6.

Sabrina Tavernise, "F.D.A. Makes It Official: BPA Can't Be Used in Baby Bottles and Cups," *The New York Times*, July 17, 2012, www.nytimes.com/2012/07/18/science/fda-bans-bpa-from-baby -bottles-and-sippy-cups.html?_r=0.

Carina Storrs, "Infant Formula Can Be a Major Source of BPA: Experts," *USNews.com*, July 6, 2012, http://health.usnews.com/health -news/news/articles/2012/07/06/infant-formula-can-be-a-major -source-of-bpa-experts.

Wendy Koch, "Fed up with BPA in Your Food? Here's How to Reduce Risks Given FDA's New 'Concerns,'" *Usatoday.com*, February 15, 2010, http://content.usatoday.com/communities/greenhouse/post /2010/01/fed-up-with-bpa-heres-how-to-ban-plastic-from-your -home/1#.UfmgRGQfZU5.

Domnique Browning, "Hitting the Bottle," *The New York Times*, May 9, 2011.

Kristin Wartman, "BPA-Free Baby Bottles Now Law, But We're Not in the Clear," *Huffingtonpost.com*, August 14, 2012, www.huffing tonpost.com/kristin-wartman/bpa-free-baby-bottles-now_b _1727752.html.

Chunyang Liao, Fang Liu, and Kurunthachalam Kannan, "Bisphenol S, a New Bisphenol Analogue, in Paper Products and Currency Bills and Its Association with Bisphenol A Residues," *Environmental Science & Technology* 46, no. 12 (May 16, 2012): 6515–22, doi:10.1021/ es300876n.

Katie Moisse, "Chemicals Banned from Toys Lurk in School Supplies," *Abcnews.go.com*, August 27, 2012, abcnews.go.com/Health/Well ness/phthalates-chemicals-banned-toys-school-supplies-center -health/story?id=17086775.

U.S. Department of Health and Human Services, "Bisphenol A (BPA) Information for Parents," Hhs.gov, www.hhs.gov/safety/bpa/.

Michelle Archer, "'Real Toy Story' Reveals Dark Side of Toy Industry," *Usatoday.com*, January 28, 2007, http://usatoday30.usatoday.com /money/books/reviews/2007-01-28-toy-usat_x.htm.

David Riley, "U.S Toy Industry Retail Sales Generated $21.18 Billion in 2011," Npd.com, January 31, 2012, www.npd.com/wps/portal /npd/us/news/press-releases/pr_120131a/.

Mary Mazzoni, "10 Back-to-School Items to Buy Used," Earth911 .com, August 16, 2011, http://earth911.com/news/2011/08/16/10 -back-to-school-items-to-buy-used/.

"Where Do Your Kids' Toys Go to Die? Children, Consumerism, Toys and Trash," *Crunchy Domestic Goddess* (blog), Crunchydomestic-goddess.com, January 19, 2010.

"FAQs: Bans on Phthalates in Children's Toys," United Stated Consumer Product Safety Commission, November 15, 2011, www.cpsc .gov/Regulations-Laws-Standards/CPSIA/Phthalates/FAQs -Bans-on-Phthalates-in-Childrens-Toys/.

Amy Westervelt, "Report Finds Toxic Levels of Phthalates Lurking in Popular Back-to-School Items," *Forbes.com*, August 27, 2012, www .forbes.com/sites/amywestervelt/2012/08/27/report-finds-toxic-lev els-of-phthalates-lurking-in-popular-back-to-school-items/.

Center for Health, Environment & Justice, *Back-to-School Guide to PVC-Free School Supplies*, 2012, http://www.chej.org/publications /PVCGuide/PVCfree2012_1.pdf.

"Loop Scoops—How We Know so Much About Juice Boxes," PBSKids .org, http://pbskids.org/loopscoops/about-juice-boxes.html.

Leah Blunt, "Recycling Mystery: Milk and Juice Cartons," Earth911 .com, 2012, http://earth911.com/news/2012/01/02/recycling-mys tery-milk-and-juice-cartons/.

Nicholas Kristof, "Warnings from a Flabby Mouse," *The New York Times*, January 20, 2013, www.nytimes.com/2013/01/20/opinion /sunday/kristof-warnings-from-a-flabby-mouse.html?smid=fb -share&_r=0&.

U.S. Department of Agriculture, *Composting Dog Waste*, 2005, ftp://
ftp-fc.sc.egov.usda.gov/AK/Publications/dogwastecomposting2.
pdf.

Janet Raloff, "Concerned About BPA: Check Your Receipts," *Science
News.org*, October 7, 2009, www.sciencenews.org/view/generic/id
/48084/description/Concerned_about_BPA_Check_your_receipts.

Jenn Savedge, "Study Finds BPA Easily Penetrates Skin," MNN.com,
November 2, 2010, www.mnn.com/family/family-activities/blogs
/study-finds-bpa-easily-penetrates-skin.

Jenn Savedge, "Which Pregnant Moms Have the Most BPA Risk
Factors?," MNN.com, October 18, 2010, http://www.mnn.com/
family/family-activities/blogs/which-pregnant-moms-have-the
-most-bpa-risk-factors.

Sonya Lunder, David Andrews, and Jane Houlihan, "BPA Coats Cash
Register Receipts," Ewg.org, July 27, 2010, www.ewg.org/bpa-in
-store-receipts.

Ryan Jaslow, "New Car Smell Is Toxic, Study Says: Which Cars Are
Worst?," *CBSNews.com*, February 15, 2012, www.cbsnews.com
/8301-504763_162-57378591-10391704/new-car-smell-is-toxic
-study-says-which-cars-are-worst/.

Jake Lingeman, "New-Car Smell Less Toxic but Not Safe, Ecology
Center Says," *Autoweek.com*, February 21, 2012, www.autoweek
.com/article/20120221/carnews/120229971.

ADDITIONAL RESOURCES

Some Other Great Books on Plastic

Plastic: A Toxic Love Story by Susan Freinkel (New York: Houghton Mifflin Harcourt, 2011).

Plastic-Free: How I Kicked the Plastic Habit and How You Can Too by Beth Terry (New York: Skyhorse Publishing, 2012).

Bottled Water

Peter Gleick, *Bottled and Sold: The Story Behind Our Obsession with Bottled Water* (Washington, DC: Island Press, 2010).

The Pacific Institute, "Bottled Water and Energy Fact Sheet," Pacinst.org, www.pacinst.org/publication/bottled-water-and-energy-a-fact-sheet/.

WeTap: find fountains near you—an app to help you find drinking water fountains and avoid the need for bottled water (http://wetap.org/).

Plastic in the Ocean

National Oceanic and Atmospheric Administration, "Plastic Marine Debris: What We Know," 2012, http://marinedebris.noaa.gov/info/plastic.html

A boat made of 12,500 plastic bottles: http://theplastiki.com/

Farmers Markets
"Farmers Markets Search," USDA.gov, 2013, http://search.ams.usda
.gov/farmersmarkets

Online Recycling Guides
Earth911, http://earth911.com/recycling/

Recycling App for Your Phone
iRecycle—available on iPhone or Android: http://earth911.com
/irecycle/

The EPA iWARM Tool: www.epa.gov/osw/conserve/tools/iwarm/
Find out exactly how much energy your recycling saves.

A Brief Guide to Non-Plastic Baby Items
Katy Balatero, "A Guide to Buying Non-plastic Baby Products,"
Grist.org, http://grist.org/article/non-plastic/

"Cloth Diapers," Cottonbabies, www.cottonbabies.com/?cPath=28

Bottle-Free Soaps and Shampoos
Lush, www.lushusa.com

Composting Pet Waste
ftp://ftp-fc.sc.egov.usda.gov/AK/Publications/dogwastecompost
ing2.pdf

INDEX

reproduction, and phthalates, 9

Resin Identification Symbol
System (RISS), 55–59

reusable bags
cloth, 129, 131–33, 188
grocery, 126, 128–32, 135

reusable bottles
BPA-free plastic, 153–54
water, 153–54

reusable coffee cups, 155–57

reusable containers, 133, 134

reuse
of newspapers and junk mail,
195–96
of plastic bags, 132–33, 135

RISS. *See* Resin Identification
Symbol System

rubber, 21–22
pacifiers, 181–82

Safeplay Systems, 72

school supplies, 184–86

sex toys, 169–71

shampoo and conditioner,
164–65

shellac, 20–21

shower curtains, 159–62

silk stockings, 30, 31

single-use plastics, 127, 190–91

six-pack rings, 140, 141–42

soaps, 163–64

Society of Plastics Engineers
(SPE), 25, 26

Society of the Plastics Industry
(SPI), 16–17, 55

soda bottles, 49, 154–55

SPE. *See* Society of Plastics
Engineers

SPI. *See* Society of the Plastics
Industry

stainless steel
bottles, 153
containers, 146–47, 188–89
ice cube trays, 146–47

stockings, 30–32

styrene, 105–6

Styrofoam, 105–6, 155

sustainability, of bioplastics,
76, 78

symbols
for purging plastic, 118–19
recycling, 55–60
RISS, 55–59

synthetic plastic, 2, 3, 18, 21, 24

synthetic polymers, 18, 45

tap water, 147–52, 154

thermoplastics, 48

thermosets, 48

toothbrushes and toothpaste,
167–69

toxins, and plastic, 93–106

toys, 182–84
sex, 169–71

Tupper, Earl Silas, 36–38, 40

Tupper Plastics, 36